Mathematical Methods in Statistics

A Workbook

by

David Freedman
University of California, Berkeley

David Lane
University of Minnesota

W. W. Norton & Company
New York London

W. W. Norton & Company, Inc., 500 Fifth Avenue, New York, N.Y. 10110
W. W. Norton & Company Ltd., 10 Coptic Street, London WC1A 1PU
ISBN 0-393-95223-1

0

Table of Contents

(Continues on next page.)

Section	Title	Page

PART B. THE ALGEBRA OF RANDOM VARIABLES

PART C. INFERENCE

Acknowledgments

Computer graphics by

Roberta Heintz and Thomas Permutt

Technical drawings by

Dale Johnson

Box art by

Georgianna Greenwood

Preface

This booklet is a supplement to the text

> STATISTICS by Freedman, Pisani and Purves
> (W.W. Norton, New York, 1978).

However, the booklet is designed so that it can be used on its own. The organization is as follows. There is a series of sections, which develop the mathematical methods and the notation in logical sequence. Each section has some explanatory material, and then a set of exercises. Mastery over the material can be gained only by working the exercises. (A pocket calculator will prove very helpful.) Answers to the odd-numbered exercises are at the end of the booklet.

The object of the first 16 sections is to present some of the mathematical reasoning commonly used in statistical work, supplementing parts II and III of the text. Topics include the average, standard deviation, correlation coefficient, and regression line. The notation may seem very abstract at first. However, this abstraction turns out to be a source of real intellectual power, for many different situations can be handled using the same equations.

The object of sections 17 through 26 is to acquaint the reader with the algebra of random variables, as commonly used in statistical work: this supplements part V of the text. The emphasis is on developing an intuitive understanding of the formalism, rather than proving theorems. However, the treatment is in a sense quite mathematical: these sections develop technique for dealing with an abstract model of chance variability. This model is applied to real examples of statistical inference only in sections 27 through 33, which cover such topics as the Gauss model and inference in simple linear regression. This supplements parts VI and VII of the text.

PART A. DESCRIPTIVE STATISTICS

1. Sigma-notation

Some algebraic notation is needed to describe the addition of a series of numbers. The Greek letter "Σ"--read "sigma"--is used for this purpose. The object of this section is to review Σ. It is easiest to begin with an example:

$$\sum_{i=1}^{3} i^2$$

This is read as follows: "summation of i^2, from 1 to 3." The Σ is an instruction to add some numbers. The subscript "$i = 1$" below Σ tells where to start the addition, at 1. The superscript "3" above Σ tells where to stop, at 3. The formula "i^2" after Σ tells what to add up, the squares. So

$$\sum_{i=1}^{3} i^2 = 1^2 + 2^2 + 3^2 = 14$$

Likewise

$$\sum_{i=1}^{4} i^2 = 1^2 + 2^2 + 3^2 + 4^2 = 30$$

$$\sum_{i=2}^{4} i^2 = 2^2 + 3^2 + 4^2 = 29$$

and

$$\sum_{i=1}^{5} i^3 = 1^3 + 2^3 + 3^3 + 4^3 + 5^3 = 225$$

The "i" is called the "index of summation". Other letters can be used. For instance

$$\sum_{j=2}^{5} j = 2 + 3 + 4 + 5 = 14$$

Exercise set 1

1. $\sum_{i=1}^{3} i = ?$

2. $\sum_{i=1}^{3} 2i = ?$

3. $\sum_{j=1}^{3} j^2 = ?$

4. $\sum_{j=1}^{3} j = ?$

5. $\sum_{j=1}^{3} (j^2 + j) = ?$

6. $\sum_{i=1}^{3} (2i + 1) = ?$

7. $\sum_{i=1}^{3} i(i + 1) = ?$

8. $\sum_{i=1}^{4} (i^2 + 1) = ?$

9. $\sum_{j=1}^{4} j(j - 1) = ?$

10. (a) $\sum_{i=1}^{4} i^2 = ?$

    (b) $\sum_{i=1}^{2} i^2 = ?$

    (c) $\sum_{i=3}^{4} i^2 = ?$

11. True or false, and explain:

$$\sum_{j=1}^{100} j^2 - \sum_{j=1}^{50} j^2 = \sum_{j=51}^{100} j^2$$

12. True or false, and explain:

$$\frac{1}{5} \times \sum_{i=1}^{5} i = \text{the average of the list } 1,2,3,4,5$$

13. One of the expressions below equals

$$\sqrt{\frac{1}{5} \times \sum_{i=1}^{5} (i-3)^2}$$

Which one, and why?

(i) $\sqrt{\frac{(-2)+(-1)+0+1+2}{5}}$

(ii) $\sqrt{\frac{(-2)^2+(-1)^2+0^2+1^2+2^2}{5}}$

(iii) $\frac{2+1+0+1+2}{5}$

14. True or false, and explain:

(a) $\sum_{j=1}^{10} j(j+1) = \sum_{k=1}^{10} k(k+1)$

(b) $\sum_{j=1}^{n} j^2 = \sum_{k=1}^{m} k^2$

15. You are given that $\sum_{j=1}^{7} j = 28$.

(a) $3 \times \sum_{j=1}^{7} j = ?$

(b) $\frac{1}{4} \times \sum_{j=1}^{7} j = ?$

16. $\sum_{j=1}^{3} 3j = ?$ Choose one option, and explain:

(i) $3+1+2+3$

(ii) $3+1+3$

(iii) $1+3+3j$

(iv) $3+6+9$

17. Express in sigma-notation:

    (a) take the integers from 1 through 17 and add them up.

    (b) take the integers from 4 through 10, square them, and add up the squares.

Terminology

The integers are the whole numbers, positive or negative, like -2, -1, 0, 1, 2.

2. Summing constants

Sigma-notation is sometimes used for adding up the same number repeatedly. An example is

(1) $$\sum_{i=1}^{3} 7$$

Here, the term to be summed is always the constant "7": it does not change with the index i. So the instruction given by the expression (1) is to take the sum of 3 sevens:

$$\sum_{i=1}^{3} 7 = 7 + 7 + 7 = 3 \times 7 = 21$$

Sometimes, this notation is used without specifying the value of the constant c:

$$\sum_{i=1}^{3} c = c + c + c = 3c$$

Likewise, the number of terms n need not be specified:

(2) $$\sum_{i=1}^{n} c = nc$$

In formula (2):

c is a constant (it does not depend on i)

n is the number of terms.

Exercise set 2

1. $\sum_{j=1}^{3} 5 = ?$

2. $\sum_{j=1}^{4} 2 = ?$

3. $\sum_{j=1}^{n} 2 = ?$

4. $\sum_{j=1}^{n} 1 = ?$

5. $\sum_{j=1}^{m} d = ?$

6. (a) $\sum_{j=1}^{5} d = ?$

   (b) $\sum_{j=1}^{3} d = ?$

   (c) $\sum_{j=4}^{5} d = ?$

7. True or false, and explain:

   (a) $\sum_{j=1}^{m} j = mj$

   (b) $\sum_{j=2}^{m} d = md$

   (c) $\sum_{j=3}^{7} d = 5d$

3. Some algebra

The object of this section is to indicate how the distributive, associative, and commutative laws of algebra apply to sigma-notation. To begin with an example, the distributive law shows that

$$2a + 2b + 2c = 2(a + b + c)$$

The same law applies when using sigma-notation. For example,

$$\sum_{i=1}^{3} 2 \times i = 2 \times \sum_{i=1}^{3} i$$

This is illustrated by exercises 1.1 and 1.2 on page 2 above.

The associative law says that it doesn't matter how you group terms when you add them up:

$$(a + b) + c = a + (b + c)$$

The commutative law says that the order of addition doesn't matter:

$$a + b = b + a$$

Here is an example where both laws are used:

$$(a + b) + (c + d) + (e + f) = (a + c + e) + (b + d + f)$$

Here is a similar example, in sigma-notation:

$$\sum_{j=1}^{3} (j^2 + j) = \left(\sum_{j=1}^{3} j^2\right) + \left(\sum_{j=1}^{3} j\right)$$

This is illustrated by exercises 1.3, 1.4 and 1.5 on page 2 above.

Exercise set 3

1. You are given that

$$\sum_{j=1}^{100} j = 5{,}050$$

and

$$\sum_{j=1}^{100} j(j-1) = 333{,}300$$

Use the distributive and associative laws to find

(a) $\sum_{j=1}^{100} 3j$

(b) $\sum_{j=1}^{100} [j(j-1) + j]$

(c) $\sum_{j=1}^{100} j^2$ Hint: $j^2 = j(j-1) + ?$

(d) $\sum_{j=1}^{100} j(j+1)$ Hint: $j(j+1) = j(j-1) + ?$

2. True or false, and explain:

(a) $\sum_{j=1}^{107} (j^2+1) = (\sum_{j=1}^{107} j^2) + 1$

(b) $\sum_{j=1}^{107} (j^2+1) = (\sum_{j=1}^{107} j^2) + 107$

(c) $\sum_{j=1}^{107} (8j)^2 = 8 \sum_{j=1}^{107} j^2$

(d) $\sum_{j=1}^{107} (8j)^2 = 64 \sum_{j=1}^{107} j^2$

3. True or false, and explain:

(a) According to the distributive law,

$$\sum_{j=1}^{100} j^2 = j(\sum_{j=1}^{100} j)$$

3. (b) According to the associative law,

$$\sum_{j=1}^{5} j^2 = (\sum_{j=1}^{5} j)^2$$

(c) $\sum_{j=1}^{100} (j^2 + 7) = (\sum_{j=1}^{100} j^2) + 700$

Notation

$a \times b$, $a \cdot b$, and ab all mean "a times b", and all three notations will be used in this booklet.

4. Statistical notation

The operation of addition is basic to statistics, as in taking averages. It should be no surprise that statisticians use sigma-notation a lot. However, some effort is needed to fit statistical data into the framework of sigma-notation. To begin with an example, a hypothetical study has 5 subjects. The letter "n" is usually reserved for the number of subjects in a study. In the example, $n = 5$. There are three variables in the study: height, weight and income. The data from the study is reported in table 1 below. Note that each subject is identified not by name, but by a code number from 1 through 5.

Table 1. Data

subject's code number	height (inches)	weight (pounds)	income (dollars)
1	69	143	17,200
2	67	139	21,400
3	71	154	28,900
4	74	148	12,400
5	67	141	9,300

The letter "x" usually stands for one of the variables in the study. In the example, x could be height, weight, or income. To get started, suppose x stands for income: or, as a statistician would say, "let x be income". Then the income of subject #3 is denoted by x_3, where the subject's code number 3 appears as a subscript on the x. The symbol "x_3" is read "eks-sub-three", or "eks-three". Keeping x for income,

$$x_5 = \text{income of subject \#5}$$
$$= \$9{,}300$$

In the next example, x stands for weight. Then

$$x_4 = \text{weight of subject \#4}$$
$$= 148 \text{ lbs}$$

Sometimes, other letters are used to stand for variables in a study. Let y stand for height. Then

$$y_2 = \text{height of subject \#2}$$
$$= 67 \text{ inches}$$

The same principles of notation can be used even with very large data sets. That is why sigma-notation can be so useful in statistics. For example, in the Health Examination Survey of 1960-62 there were 6,672 subjects.[1] So $n = 6{,}672$. The subjects are identified only by code-numbers from 1 to 6,672, protecting the confidentiality of their responses to the survey. Over two hundred variables were collected for each subject--including height, weight and income. Let x stand for height. Then $x_{2,136}$ stands for the height of the subject with code number 2,136 in Health Examination Survey. As it happens, subject #2,136 was female, aged 23, married, and living in a big city. She was 67 inches tall, weighed 131 lbs and her systolic blood pressure was 103 mm. So $x_{2,136} = 67$ inches.

To conclude this section, statistical notation will be used to summarize the discussion in section 3. The letter i will be used to stand

[1] See page 49 of the text.

for the code-number of a subject. For example, let x be height. Then x_i stands for the height of the subject with code number i. And $\sum_{i=1}^{n} x_i$ stands for the sum of the heights of all the subjects in the study.

In this notation, the distributive law can be written as follows:

$$(1) \qquad \sum_{i=1}^{n} cx_i = c \sum_{i=1}^{n} x_i$$

The associative and commutative laws show

$$(2) \qquad \sum_{i=1}^{n} (y_i + x_i) = \sum_{i=1}^{n} y_i + \sum_{i=1}^{n} x_i$$

There is one monster formula, combining (1) and (2), as well as formula 2.2 on page 5:

$$(3) \qquad \sum_{i=1}^{n} (by_i + cx_i + d) = (b \sum_{i=1}^{n} y_i) + (c \sum_{i=1}^{n} x_i) + (nd)$$

In these formulas, b, c and d are constants: they do not depend on the index of summation i. Equation (3) can be extended to three or more variables.

Exercise set 4

1. One month, the Gallup poll interviewed 2,438 subjects.[1] So $n = ?$

2. Another month, the Gallup poll interviewed 1,397 subjects. So $n = ?$

[1] See chapter 19 of the text.

3. Let x stand for blood pressure, in the Health Examination Survey of 1960-62. Then $x_{2,136} = ?$ (This information is given in the reading material above.)

4. Suppose y stands for height. Then, for Table 1 on page 10 above, $y_3 = ?$

5. Suppose z stands for weight. Then, for Table 1 on page 10 above, $z_2 = ?$

6. Suppose x stands for income. Then, for Table 1 on page 10 above:

   (a) $x_1 = ?$

   (b) $\sum_{i=1}^{5} x_i = ?$

   (c) $\frac{1}{5} \sum_{i=1}^{5} x_i = ?$

   (d) $\sum_{i=1}^{3} x_i^2 = ?$

7. The average income for the 5 subjects in Table 1 on page 10 above was ____________.

8. Write out the incomes of the subjects in Table 1 on page 10 above, as a list of 5 numbers. What is the average of this list?

9. Suppose you list out the values of a variable for all the subjects in a study. True or false, and explain:

   (a) The number of entries on the list will be n.

   (b) The average of the list will equal the average of the variable.

10. Let x be income, in the Health Examination Survey. True or false, and explain:

(a) $\sum_{i=1}^{6,672} x_i$ is the total income of all the subjects in the study.

(b) $\sum_{i=1}^{6,672} x_i^2$ is the sum of the squared incomes, for all the subjects in the study.

(c) The units for $\sum_{i=1}^{6,672} x_i$ are dollars.

(d) The units for $\sum_{i=1}^{6,672} x_i^2$ are square dollars.

11. True or false, and explain:

(a) $\sum_{j=1}^{138} (3u_j - 5v_j + 1) = 3 \sum_{j=1}^{138} u_j - 5 \sum_{j=1}^{138} v_j + 138$

(b) $\sum_{j=1}^{m} u_j + \sum_{k=1}^{n} v_k = \sum_{i=1}^{m+n} (u_i + v_i)$

12. A marketing research firm interviewed 250 people in a shopping mall. Each person interviewed was assigned a code number, corresponding to the order in which he or she was interviewed. The interviewers asked how much cash each subject was carrying, in bills and in change. Let x be the amount of cash in bills, and y the amount of cash in change. Write the statistical notation for the following quantities:

(a) the amount of cash in bills carried by the seventh person interviewed;

(b) the total amount of cash, in bills and change, carried by the thirtieth person interviewed;

(c) the total amount of cash in change carried by all the people interviewed;

(d) the average amount of change carried by the first 100 people interviewed.

13. Let x and y be the same as in exercise 12. What are the units of the following quantities?

(a) x_{10}

(b) $\sum_{i=1}^{250} x_i$

(c) $\frac{1}{250} \sum_{i=1}^{250} x_i$

(d) x_{10}^2

(e) $\sum_{i=1}^{250} x_i^2$

(f) $\frac{1}{250} \sum_{i=1}^{250} x_i^2$

14. For this exercise, x and y are variables in a study, c and d are constants. True or false, and explain:

(a) $\sum_{j=1}^{J} cy_j = c \sum_{j=1}^{J} y_j$

(b) $\sum_{i=1}^{n} (x_i + c) = (\sum_{i=1}^{n} x_i) + (nc)$

(c) $\sum_{j=1}^{m} (y_j + d) = (\sum_{j=1}^{m} y_j) + (nd)$

(d) $\sum_{i=1}^{n} (x_i - c) = (\sum_{i=1}^{n} x_i) - c$

[Continues on next page.]

(e) $\sum_{i=1}^{m} (x_i + y_i) = \sum_{i=1}^{m} x_i + \sum_{i=1}^{m} y_i$

(f) $\sum_{i=1}^{n} (x_i + c)^2 = (\sum_{i=1}^{n} x_i^2) + 2c(\sum_{i=1}^{n} x_i) + nc^2$

Hint: $(x_i + c)^2 = x_i^2 + 2cx_i + c^2$

(g) $\sum_{i=1}^{n} (x_i + y_i)^2 = (\sum_{i=1}^{n} x_i^2) + 2(\sum_{i=1}^{n} x_i)(\sum_{i=1}^{n} y_i) + (\sum_{i=1}^{n} y_i^2)$

(h) $\sum_{i=1}^{n} (x_i + y_i)^2 = (\sum_{i=1}^{n} x_i^2) + 2(\sum_{i=1}^{n} x_i y_i) + (\sum_{i=1}^{n} y_i^2)$

15. By accident, all 7 subjects in a study of college students happen to have the same age. They are all nineteen. Their average age is ____.

16. Let u and v be variables in a study with m subjects. True or false, and explain:

$$\sum_{j=1}^{m} (12u_j - 17v_j + 5) = (12 \sum_{j=1}^{m} u_j) - (17 \sum_{j=1}^{m} v_j) + (5m)$$

Notation

Sometimes, other letters are used to denote the number of subjects in a study: m and N are quite common.

5. The average

We can now use sigma-notation to give a mathematical formula for the average, and to prove certain basic facts about the average. In the formula, n stands for the number of subjects in a study. These subjects are given code-numbers from 1 through n. The letter "x" stands for some variable in the study, for instance, height. Then x_i is the height of the subject with code number i. And $\sum_{i=1}^{n} x_i$ represents the sum of the heights of all the subjects in the study. To obtain the average, it is necessary to divide this sum by n, the number of subjects. In mathematical notation, then, the average is

$$\frac{1}{n} \sum_{i=1}^{n} x_i$$

The average is usually denoted by a bar. So

$$\bar{x} = \frac{1}{n} \sum_{i=1}^{n} x_i \tag{1}$$

Formula (1) expresses, in mathematical notation, the fact that the average of a list is obtained by adding up all the numbers on the list, and dividing by how many there are. The units for $\bar{x}$ are, of course, the same as the units for x. Average height comes out in inches, average weight in pounds. It is a good idea to keep track of units. For the connection between variables and lists, see exercises 4.7-4.9 on page 13 above.

We will now establish some facts about the average. To begin with, if the values of a variable are the same, the average is just that common value. To state this as a formula, let c denote the common value.

$$(2) \qquad \text{If } x_i = c \text{ for all } i, \text{ then } \bar{x} = c .$$

The proof of (2) starts from formula 2.2 on page 5 above, which says that

$$(3) \qquad \sum_{i=1}^{n} c = nc$$

Dividing both sides by n gives formula (2).

Next, suppose we multiply all the values of a variable by the same number c. This just multiplies the average by c. To put this as a formula, let $z_i = cx_i$ for all i: In other words, for every subject in the study, the value of z is just the constant c times the value of x.

$$(4) \qquad \text{If } z_i = cx_i \text{ for all } i, \text{ then } \bar{z} = c\bar{x} .$$

This follows from the distributive law: formula 4.1 on page 12 above.

Next, suppose the same number d is added to all the values. That adds d to the average. The formula:

$$(5) \qquad \text{If } z_i = x_i + d \text{ for all } i, \text{ then } \bar{z} = \bar{x} + d .$$

Here is a proof. By formula 4.2 on page 12 above,

$$(6) \qquad \frac{1}{n} \sum_{i=1}^{n} (x_i + d) = \frac{1}{n}[(\sum_{i=1}^{n} x_i) + (\sum_{i=1}^{n} d)]$$

Now use formula (3), with d in place of c:

$$\begin{aligned}\bar{z} &= \frac{1}{n}\sum_{i=1}^{n}(x_i + d)\\ &= \frac{1}{n}[(\sum_{i=1}^{n} x_i) + (nd)]\\ &= (\frac{1}{n}\sum_{i=1}^{n} x_i) + d\\ &= \bar{x} + d\end{aligned}$$

This completes the argument for (5).

Along these lines, formula 4.3 on page 12 above can be restated in terms of averages:

(7) Proposition. Let x and y be two variables in a study. Let b, c and d be three constants. Define a new variable z as follows: $z_i = by_i + cx_i + d$ for all subjects i. Then $\bar{z} = b\bar{y} + c\bar{x} + d$.

This combines (4) and (5); it can be extended to three or more variables.

The last topic in this section is the deviations from average. Let x be a variable in a study, with average $\bar{x}$. How far is subject #1 from average? The answer is $x_1 - \bar{x}$. For subject #2, the deviation is $x_2 - \bar{x}$. And so on. These are the deviations from average. For subject #i, the deviation from average is $x_i - \bar{x}$.

We will now prove a basic fact about the deviations from average: that their average is 0. This is necessarily so, for any variable in any study. This fact requires a mathematical proof: checking some examples helps, but does not prove that the average deviation will be 0 in every case. The proof will show how sigma-notation is used.

(8) Proposition. For any variable, the average of the deviations from average equals 0. As a formula,

$$\frac{1}{n}\sum_{i=1}^{n}(x_i - \bar{x}) = 0$$

Proof. Use formula (5), substituting $-\bar{x}$ for d: this may look a bit funny, but it is legitimate because $\bar{x}$ does not depend on i:

$$\frac{1}{n}\sum_{i=1}^{n}(x_i - \bar{x}) = \bar{x} - \bar{x} = 0$$

This completes the proof.

Exercise set 5

1. Five men in a study were examined in a hospital. Their heights in inches were:

   68, 73, 66, 68, 70

   Their average height is

$$\frac{68 + 73 + 66 + 68 + 70}{5} = 69 \text{ inches}$$

   Person A wishes to find the average height in centimeters by the following procedure:

   (A) multiply each height on the list by 2.54 cm/in, to convert to cm, and then take the average of the new list

[Continues on next page.]

Person B wants to use another procedure:

(B) multiply the average, 69 inches, by 2.54 cm/in

Which is right? Or perhaps both are? Or neither? Explain briefly.

2. Find the deviation from average height for each man in exercise 1. Compute the average of these deviations.

3. The five men in exercise 1 were weighed too, and their average weight was 147 lbs. However, they were all wearing standard hospital clothing during the examination. This clothing weighs 2 lbs. Their average body weight was ________. Explain briefly.

4. True or false, and explain: if $n = 3$, the average of x is

$$\frac{x_1 + x_2 + x_3}{3}$$

5. In Table 1 on page 10 above, find the average of the squared heights of the subjects. Find the square of the average height. Are these the same? (Note: Your units must be square inches.)

6. In Table 1 on page 10 above, let x stand for height. Find

$$\frac{1}{5} \sum_{i=1}^{5} x_i^2 \quad \text{and} \quad \left(\frac{1}{5} \sum_{i=1}^{5} x_i\right)^2$$

Are these the same?

7. The table below is a partial report of data from the Health Examination Survey.

(a) If x stands for height and $i = 1$, find x_i.

(b) If x stands for weight and $j = 1{,}420$, find x_j.

(c) If x stands for blood pressure and $k = 6{,}672$, find x_k.

(d) If x stands for age and $x_i = 43$ years, find i.

Table 1. Data from the Health Examination Survey

Code number	Sex	Age (years)	Height (inches)	Weight (pounds)	Blood pressure (mm Hg)
1	Female	45	68	148	113
872	Male	64	64	136	147
1,420	Female	45	61	156	138
2,136	Female	23	67	131	103
4,131	Female	43	62	176	113
6,672	Female	22	60	112	118

8. $\sum_{i=1}^{3} x_1 = x_1 + x_2 + x_3$? or $3x_1$? Explain briefly.

9. $\sum_{i=1}^{5} x_2 = x_1 + x_2 + x_3 + x_4 + x_5$? or $5x_2$? Explain briefly.

10. True or false, and explain:

(a) $(\sum_{i=1}^{3} x_i)^2 = \sum_{i=1}^{3} x_i^2$

(b) $(\frac{1}{3} \sum_{i=1}^{3} x_i)^2 = \frac{1}{3} \sum_{i=1}^{3} x_i^2$

11. True or false, and explain:

$$\frac{1}{N} \sum_{n=1}^{N} (3u_n - \tfrac{1}{2}v_n + \tfrac{5}{8}) = (3\frac{1}{N}\sum_{n=1}^{N} u_n) - (\frac{1}{2}\frac{1}{N}\sum_{n=1}^{N} v_n) + \frac{5}{8}$$

12. Let z stand for income, in the Health Examination Survey. The average is

$$\bar{z} = \frac{1}{6,672} \sum_{k=1}^{6,672} z_k$$

(a) Copy out the part of the formula which represents the income of subject #k in this study.

(b) Copy out the part of the formula which represents the total income of all the subjects in this study.

(c) What does $z_k - \bar{z}$ represent, in words?

(d) Find the average of $z_k - \bar{z}$. Or can this be done without the data?

13. In a report on a study, the following formula occurred:

$$\bar{y} = \frac{1}{m} \sum_{k=1}^{m} y_k$$

(a) In this formula, what does the bar over the y denote?

(b) Which symbol in the formula stands for the number of subjects?

(c) Which symbol stands for the variable?

(d) What does the symbol "k" stand for?

14. Two surveyors are working on the lots in a subdivision. One measures the lengths; the other does the widths. Due to a misunderstanding, the first one works in inches, and the second in centimeters. There are 110 lots in all. Their average length is 1207 inches. Their average width is 1524 centimeters. [Continues on next page.]

(a) If possible, find the average perimeter of the lots, in feet.

(b) If possible, find the average area of the lots, in square feet.

Explain carefully.

Note. 1 foot = 12 inches, and 1 inch = 2.54 centimeters.

15. True or false, and explain: if $z_i = cx_i + d$ for all i, then $\bar{z} = c\bar{x} + d$.

16. True or false, and explain:

(a) If u is a variable in a study with N subjects, then $\bar{u} = \sum_{n=1}^{N} (u_n/N)$

(b) If v is a variable in a study with M subjects, then $\bar{v} = \sum_{n=1}^{M} (v_n/n)$

(c) If w is a variable in a study with n subjects, then $\frac{1}{n} \sum_{i=1}^{n} \bar{w}^2 = \bar{w}^2$.

17. If $a \leqq x_i \leqq b$ for all subjects i, prove that $a \leqq \bar{x} \leqq b$.

Notation

$u \leqq v$ means that u is less than or equal to v.

In formula (7), we had two variables, x and y, and three constants, b, c and d. We defined a new variable z as follows: $z_i = by_i + cx_i + d$ for all subjects i. This is often abbreviated as follows:

$$z = by + cx + d$$

6. The standard deviation

The object of this section is to express the standard deviation in sigma-notation, and to establish some of its properties. To review briefly, n is the number of subjects in a study, x stands for a variable, and x_i is the value for subject #i in the study. The average is

$$\bar{x} = \frac{1}{n} \sum_{i=1}^{n} x_i$$

Consequently, the deviation of x_1 from the average is $x_1 - \bar{x}$; the deviation of x_2 from the average is $x_2 - \bar{x}$; and so on, up to x_n. These deviations are

$$x_1 - \bar{x},\ x_2 - \bar{x}, \ldots, x_n - \bar{x}$$

The standard deviation is the root-mean-square[1] of these deviations:

$$\sqrt{\frac{1}{n} \sum_{i=1}^{n} (x_i - \bar{x})^2}$$

"Root-mean-square" is best read backwards, like a German sentence:

Step 1. <u>Square</u> the deviations.

Step 2. Take the <u>mean</u> (or average) of the squares.

Step 3. Take the square <u>root</u> of the mean.

By convention, the square root is never negative. Root-mean-square is often abbreviated to "r.m.s.", and the standard deviation to "SD".

The units for the standard deviation are the same as the units for the variable. The standard deviation of income comes out in dollars, the

[1] See page 57 of the text.

standard deviation of height, in inches. Usually, the standard deviation is denoted by the letter "s". So the formula for the standard deviation is

$$(1) \qquad s = \sqrt{\frac{1}{n} \sum_{i=1}^{n} (x_i - \bar{x})^2}$$

The formula illustrates two principles of algebra:

(i) Symbols stand for numbers. In (1), the number $\bar{x}$ is to be subtracted from the number x_i.

(ii) It helps to have one symbol to stand for a complicated expression. In (1), the symbol $\bar{x}$ stands for $\frac{1}{n} \sum_{j=1}^{n} x_j$. You may not like (1), but it is certainly easier to read than

$$\sqrt{\frac{1}{n} \sum_{i=1}^{n} [x_i - (\frac{1}{n} \sum_{j=1}^{n} x_j)]^2}$$

We can now prove some facts about the standard deviation. The first is that multiplying every value of a variable by the same positive number c just multiplies the SD of the variable by c.

(2) Proposition. Let $z_i = cx_i$ for all subjects i, where c is a positive constant. Then the standard deviation of z equals c times the standard deviation of x.

Proof. As formula 5.4 on page 18 above shows, $\bar{z} = c\bar{x}$. So

$$z_i - \bar{z} = c(x_i - \bar{x})$$

Now

$$\frac{1}{n} \sum_{i=1}^{n} (z_i - \bar{z})^2 = \frac{1}{n} \sum_{i=1}^{n} [c(x_i - \bar{x})]^2$$

$$= \frac{1}{n} \sum_{i=1}^{n} c^2(x_i - \bar{x})^2$$

So

$$\frac{1}{n}\sum_{i=1}^{n}(z_i-\bar{z})^2 = c^2\,\frac{1}{n}\sum_{i=1}^{n}(x_i-\bar{x})^2$$

Using the principle $\sqrt{ab} = \sqrt{a}\cdot\sqrt{b}$,

$$\sqrt{\frac{1}{n}\sum_{i=1}^{n}(z_i-\bar{z})^2} = \sqrt{c^2}\,\sqrt{\frac{1}{n}\sum_{i=1}^{n}(x_i-\bar{x})^2}$$

$$= c\,\sqrt{\frac{1}{n}\sum_{i=1}^{n}(x_i-\bar{x})^2}$$

This completes the proof.

Next, we will prove that adding the same number to every value of a variable does not change the SD.

(3) Proposition. Let $z_i = x_i + d$ for all subjects i . Then z and x have the same standard deviation.

Proof. As formula 5.5 on page 18 above shows, $\bar{z} = \bar{x} + d$. Now

$$z_i - \bar{z} = (x_i + d) - (\bar{x} + d) = x_i - \bar{x}$$

So

$$\sqrt{\frac{1}{n}\sum_{i=1}^{n}(z_i-\bar{z})^2} = \sqrt{\frac{1}{n}\sum_{i=1}^{n}(x_i-\bar{x})^2}$$

The left side of this equality is the standard deviation of z: the right side is the standard deviation of x. This completes the proof.

The next proposition justifies an alternative procedure for computing the standard deviation.[1]

[1] See page 65 of the text.

(4) Proposition. $\sqrt{\frac{1}{n}\sum_{i=1}^{n}(x_i - \bar{x})^2} = \sqrt{(\frac{1}{n}\sum_{i=1}^{n} x_i^2) - (\bar{x}^2)}$

Proof. First, $(x_i - \bar{x})^2 = x_i^2 - 2\bar{x}x_i + \bar{x}^2$. Now use proposition 5.7 on page 19 above:

$$(5) \qquad \frac{1}{n}\sum_{i=1}^{n}(x_i - \bar{x})^2 = \frac{1}{n}\sum_{i=1}^{n}(x_i^2 - 2\bar{x}x_i + \bar{x}^2)$$

$$= (\frac{1}{n}\sum_{i=1}^{n} x_i^2) - 2\bar{x}(\frac{1}{n}\sum_{i=1}^{n} x_i) + \bar{x}^2$$

$$= (\frac{1}{n}\sum_{i=1}^{n} x_i^2) - 2\bar{x}^2 + \bar{x}^2$$

$$= (\frac{1}{n}\sum_{i=1}^{n} x_i^2) - \bar{x}^2$$

This is legitimate because $\bar{x}$ and $\bar{x}^2$ are constants: they do not depend on the index of summation i . In proposition 5.7, we are substituting

1 for b and x_i^2 for y_i

$-2\bar{x}$ for c and $\bar{x}^2$ for d

The conclusion is:

$$(6) \qquad \frac{1}{n}\sum_{i=1}^{n}(x_i - \bar{x})^2 = (\frac{1}{n}\sum_{i=1}^{n} x_i^2) - \bar{x}^2$$

Taking the square root of both sides of (5) completes the proof.

The alternative formula for the standard deviation is

$$(7) \qquad s = \sqrt{(\frac{1}{n}\sum_{i=1}^{n} x_i^2) - (\bar{x}^2)}$$

In words: the standard deviation is the square root of the average of the squares, minus the square of the average.

Exercise set 6

1. True or false, and explain.

   (a) If $y_i = -x_i$ for all subjects i, then x and y have the same average.

   (b) If $y_i = -x_i$ for all subjects i, then x and y have the same standard deviation.

2. There were 411 men aged 18 to 24 in the Health Examination Survey of 1960. The sum of their heights was 28,213 inches. The sum of the squares of their height was 1,939,325 square inches. If possible, find the average and SD of their heights from this information. If this is impossible, explain why.

3. Let $n = 3$ and $m = \frac{x_1 + x_2 + x_3}{3}$. True or false, and explain:

   (a) The standard deviation of x equals

$$\sqrt{\frac{(x_1 - m)^2 + (x_2 - m)^2 + (x_3 - m)^2}{3}}$$

   (b) The standard deviation of x equals

$$\sqrt{\frac{x_1^2 + x_2^2 + x_3^2}{3} - m^2}$$

4. In one United States study, the body temperatures of 387 students were measured in degrees Fahrenheit: the average was 98.6 degrees, with an SD of 0.2 degrees. For presentation at a European conference, all 387 temperatures were converted to degrees Centigrade:

$$\text{Centigrade} = \frac{5}{9}(\text{Fahrenheit} - 32)$$

[Continues on next page.]

From the information given, can you find the average and SD of the converted temperatures? If so, what are they? If not, why not?

5. True or false, and explain:

(a) The standard deviation is always 0 or positive.

(b) If the standard deviation of x is 0, then $x_i = \bar{x}$ for all i.

(c) If the standard deviation of x is 0, then $x_i = 0$ for all i.

(d) If $x_i = c$ for all i, then the standard deviation of x is 0.

(e) If the average of x is 0, then $x_i = 0$ for all i.

6. Calculate the SD of the list 1,2,3,4,5 by the two methods of this section.

7. Let y be systolic blood pressure in the Health Examination Survey. The standard deviation of y is

$$\sqrt{\frac{1}{6,672}\sum_{j=1}^{6,672} y_j^2 - \left(\frac{1}{6,672}\sum_{j=1}^{6,672} y_j\right)^2}$$

(a) Copy out the part of the formula which gives the blood pressure of subject #j. Or does this come into the formula?

(b) Copy out the part of the formula which gives the average blood pressure of the subjects. Or does this come into the formula?

(c) Copy out the part of the formula which gives the sum of the squares of the blood pressures for all the subjects. Or does this come into the formula?

(d) Copy out the part of the formula which gives the deviation from average for subject #j. Or does this come into the formula?

8. Let $z_i = cx_i + d$, for all i. Prove that

$$\text{SD of } z = |c| \cdot \text{SD of } x$$

<u>Notation</u>: $|c|$ is the <u>absolute</u> <u>value</u> of c;

$$|c| = c \quad \text{if} \quad c > 0$$
$$|c| = 0 \quad \text{if} \quad c = 0$$
$$|c| = -c \quad \text{if} \quad c < 0$$

9. This exercise is on <u>standard</u> <u>units</u>.[1] To convert the value of a variable to standard units, subtract the average and divide by the SD. This says how many SDs above (+) or below (-) the average this value is. The variable x has average $\bar{x}$ and standard deviation s. So, in standard units,

$$x_1 \text{ becomes } (x_1 - \bar{x})/s$$
$$x_2 \text{ becomes } (x_2 - \bar{x})/s$$
$$\vdots$$
$$x_n \text{ becomes } (x_n - \bar{x})/s$$

Let z_i be x_i in standard units:

$$z_i = (x_i - \bar{x})/s$$

(a) Prove that $\bar{z} = 0$.

(b) Prove that the standard deviation of z equals 1.

In words: When converted to standard units, a variable has average equal to 0 and SD equal to 1.

[1] See page 71 of the text. It is tacitly assumed that $s > 0$.

(c) If x stands for weight in pounds, what are the units for z?

(d) If x stands for weight in kilograms, what are the units for z?

10. Let $u_i = cx_i + d$. True or false, and explain: If $c > 0$, there is no difference between u in standard units and x in standard units. What if $c < 0$?

11. In Table 1 on page 10 above, let x stand for height and y for weight.

(a) If $i = 3$, what are x_i, y_i, and $x_i \cdot y_i$? (Give units.)

(b) Find $\frac{1}{5}\sum_{i=1}^{5} x_i \cdot y_i$. How does this compare with $\bar{x} \cdot \bar{y}$?

12. For any variable z and any constant d prove that

$$\frac{1}{n}\sum_{i=1}^{n} (z_i - d)^2 = [\frac{1}{n}\sum_{i=1}^{n} (z_i - \bar{z})^2] + (\bar{z} - d)^2$$

Hint: Imitate the proof of (4), starting with the fact

$$\begin{aligned}(z_i - d)^2 &= [(z_i - \bar{z}) + (\bar{z} - d)]^2 \\ &= (z_i - \bar{z})^2 + 2(\bar{z} - d)(z_i - \bar{z}) + (\bar{z} - d)^2\end{aligned}$$

13. By definition, the root-mean-square deviation of x from d is

$$\sqrt{\frac{1}{n}\sum_{i=1}^{n} (x_i - d)^2}$$

Prove that this is smallest when $d = \bar{x}$. What is this smallest value?

Hint: Use exercise 12.

14. If $n = 2$, prove that SD of $x = \frac{1}{2}|x_1 - x_2|$.

15. If n is even and $a \leqq x_i \leqq b$ for all i, how big can the SD of x be? (Hard.)

7. Variance

The next important object of study is the normal curve. However, it is convenient to pause here, and introduce variance. Let x stand for a variable in a study with n subjects. These subjects have code numbers i ranging from 1 to n. The variance of x is defined as

$$\frac{1}{n} \sum_{i=1}^{n} (x_i - \bar{x})^2$$

In words, the variance of x is the average square deviation of x from its average.

It is important to keep track of units in the variance, although they are funny: they are the square of the units for x. The variance of height comes out in square inches; the variance of weight comes out in square pounds.

The variance of x is usually abbreviated as "var x". The procedure for computing variance can be expressed as a formula:

$$\text{(1)} \qquad \text{var } x = \frac{1}{n} \sum_{i=1}^{n} (x_i - \bar{x})^2$$

Comparing this with formula 6.1 on page 26 above, you can recognize var x as the quantity inside the square root sign of 6.1 :

$$\text{(2)} \qquad \text{SD of } x = \sqrt{\text{var } x}$$

Formula 6.6 on page 28 above gives an alternative way to compute the variance:

$$\text{(3)} \qquad \text{var } x = \left(\frac{1}{n}\sum_{i=1}^{n} x_i^2\right) - \bar{x}^2$$

In words, the variance of x is the average of the squares, minus the square of the average.

There is still another formula for var x, which looks at first glance like a typographical error:

$$\text{(4)} \qquad \text{var } x = \frac{1}{n}\sum_{i=1}^{n} x_i(x_i - \bar{x})$$

This is so because

$$\begin{aligned}\text{var } x &= \frac{1}{n}\sum_{i=1}^{n} (x_i - \bar{x})^2\\ &= \frac{1}{n}\sum_{i=1}^{n} (x_i - \bar{x})\cdot(x_i - \bar{x})\\ &= \frac{1}{n}\sum_{i}^{n} [x_i(x_i - \bar{x}) - \bar{x}(x_i - \bar{x})]\end{aligned}$$

But $\bar{x}$ is a constant, so formula 5.7 on page 19 applies:

$$\text{var } x = \left[\frac{1}{n}\sum_{i=1}^{n} x_i(x_i - \bar{x})\right] - \bar{x}\left[\frac{1}{n}\sum_{i=1}^{n} (x_i - \bar{x})\right]$$

And $\frac{1}{n}\sum_{i=1}^{n} (x_i - \bar{x}) = 0$ by proposition 5.8 on page 20 above. This completes the argument for (4).

Variance is an important technical concept, but without any direct intuitive meaning. To interpret a variance, take its square root. This gives the SD, which measures spread around the average.

Exercise set 7

1. True or false, and explain:

   (a) variance = SD^2

   (b) variance = $\sqrt{SD}$

   (c) SD = $\sqrt{\text{variance}}$

   (d) SD = variance^2

2. Compute the variance of the height data in Table 1 on page 10 above. Be sure to give units in the answer.

3. For a certain group of children, the variable x is defined as height in centimeters: $\bar{x} = 150$ and $\text{var } x = 20$. A typical child in this group is around 150 cm tall, give or take

   (i) 20 cm or so

   or (ii) 4.5 cm or so.

   Choose one option, and explain why.

4. For the same group of children, weight is measured in kilograms. The variance of weight is

   (i) 9 kg

   or (ii) 81 kg^2

   Choose one option and explain why.

5. For a certain group of women, the standard deviation of years of schooling completed was 3 years. The variance of this variable was ____________. Be sure to give units.

6. For this same group of women, the variance of family incomes was 100,000,000 dollars^2. The standard deviation of this variable was __________. Be sure to give units. Or is something wrong with the data?

7. Let $z_i = cx_i + d$. Prove that $\text{var } z = c^2 \text{ var } x$. How is this related to exercise 6.8 on page 31?

8. True or false, and explain: $\text{var } x = \frac{1}{n} \sum_{i=1}^{n} (x_i - \bar{x})x_i$.

9. True or false, and explain: doubling every value of a variable doubles the variance.

10. True or false, and explain: adding three to every value of a variable adds three to the variance.

11. Under what circumstances is the average of the squares equal to the square of the average?

12. True or false, and explain:

(a) $\sqrt{\frac{1}{n} \sum_{i=1}^{n} (x_i - \bar{x})^2} = \frac{1}{n} \sum_{i=1}^{n} |x_i - \bar{x}|$

(b) $\frac{1}{n} \sum_{i=1}^{n} (x_i^3 + \bar{x}^2) = (\frac{1}{n} \sum_{i=1}^{n} x_i^3) + (\bar{x}^2)$

(c) $\frac{1}{n} \sum_{i=1}^{n} x_i^2 = (\frac{1}{n} \sum_{i=1}^{n} x_i)^2$

8. The normal curve

For many variables, the normal approximation to data applies: the histogram of the variable put into standard units is closely approximated by the standard normal curve.[1] Sometimes, it is useful to approximate the histogram, with the variable expressed in the original units. For this purpose, statisticians use a whole family of normal curves. These curves are identified by two numbers, μ and σ^2. These are greek letters: μ is read "mu", and stands for the mean (or average); σ is read "sigma", and represents the standard deviation. Thus, σ^2 represents the variance. The normal curve with mean μ and variance σ^2 is often abbreviated by $N(\mu,\sigma^2)$, which is read "normal mu sigma-squared" or, more briefly, "enn mu sigma-squared".

The standard normal curve has the equation

$$(1) \qquad f(z) = 100\% \cdot \frac{1}{\sqrt{2\pi}} \cdot e^{-\frac{1}{2}z^2}$$

In this equation, z represents a variable in standard units. The density $f(z)$ represents the height of an idealized (smoothed) histogram for this variable.[2] The units for $f(z)$ are: percent per standard unit.

The equation for the $N(\mu,\sigma^2)$ curve is

$$(2) \qquad g(x) = \frac{1}{\sigma} \cdot 100\% \cdot \frac{1}{\sqrt{2\pi}} \cdot e^{-\frac{1}{2}\left(\frac{x-\mu}{\sigma}\right)^2}$$

[1] See chapter 5 of the text.

[2] Densities and histograms are covered in chapter 3 of the text.

In this equation:

μ represents the average

σ represents the standard deviation

x is a typical value of the variable

$\frac{x-\mu}{\sigma}$ is this value converted to standard units

$g(x)$ is the density (of the idealized histogram) at x

Although the standard deviation σ is the basic quantity in (2), it is the variance σ^2 which is used in the notation $N(\mu,\sigma^2)$. There is no help for it, every statistician uses this notation.

Figure 1. The normal curve, drawn in standard units and in inches.

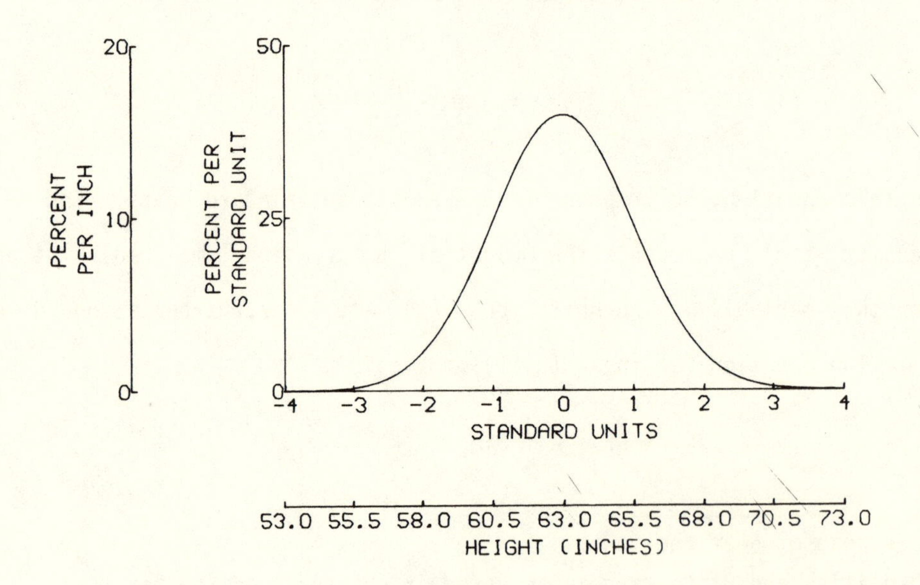

At this point, an example may be useful. Women's heights follow the normal curve, with an average of 63 inches and an SD of 2.5 inches:

$$\mu = 63 \text{ inches} \quad \text{and} \quad \sigma = 2.5 \text{ inches}.$$

Figure 1 on page 38 shows the normal curve, both in standard units and in original units--inches. In standard units, this is the standard normal curve. In original units, this is the $N(63, 2.5^2)$ curve. However, as the figure shows, the two curves are exactly the same, apart from the units. The inside horizontal and vertical scales are for standard units. The outside horizontal and vertical scales are for inches.

What makes it all match up this way? Take the top of the inner axis. This is 50% per standard unit. But each standard unit covers 2.5 inches. Spreading 50% over a standard unit is like spreading 50% over 2.5 inches, giving $50/2.5 = 20\%$ per inch. So the top of the inner axis matches the top of the outer axis. Any other pair of values can be dealt with the same way. In general, P% per standard unit matches $\frac{1}{\sigma}$ P% per original unit; this may help to explain why the $\frac{1}{\sigma}$ is needed on the right side of (2).

For readers with calculus, the $1/\sqrt{2\pi}$ is needed to make the total area under the normal curve work out to 1, or 100%. It is a mathematical fact that

$$\int_{-\infty}^{\infty} e^{-\frac{1}{2}z^2}\,dz = \sqrt{2\pi} \tag{3}$$

So

$$\frac{1}{\sqrt{2\pi}}\int_{-\infty}^{\infty} e^{-\frac{1}{2}z^2}\,dz = 1 \tag{4}$$

In original units, the factor $\frac{1}{\sigma}$ makes the total area under the $N(\mu,\sigma^2)$ curve come out to 1. Indeed,

$$(5) \qquad \frac{1}{\sigma}\frac{1}{\sqrt{2\pi}}\int_{-\infty}^{\infty} e^{-\frac{1}{2}\left(\frac{x-\mu}{\sigma}\right)^2} dx = \frac{1}{\sqrt{2\pi}}\int_{-\infty}^{\infty} e^{-\frac{1}{2}z^2} dz = 1$$

The first equality in (5) can be verified by making the change of variables

$$z = \frac{x-\mu}{\sigma}$$

The second equality in (5) is just (4).

When drawn in original units rather than standard units, the normal curves can look quite different. For example, men's heights follow the normal curve with mean 69 inches and SD 3 inches. This is the $N(69,3^2)$-curve. Figure 2 compares the solid $N(69,3^2)$-curve with the dashed $N(63,2.5^2)$-curve for women, both drawn in original units--inches. The men's curve is farther to the right, because the mean is larger: 69 compared to 63. The men's curve is also flatter and more spread out, because the SD is larger: 3 compared to 2.5. Being flatter, it has to spread out more, to make the total areas under the two curves equal to 1, or 100%.

Figure 2. Two normal curves

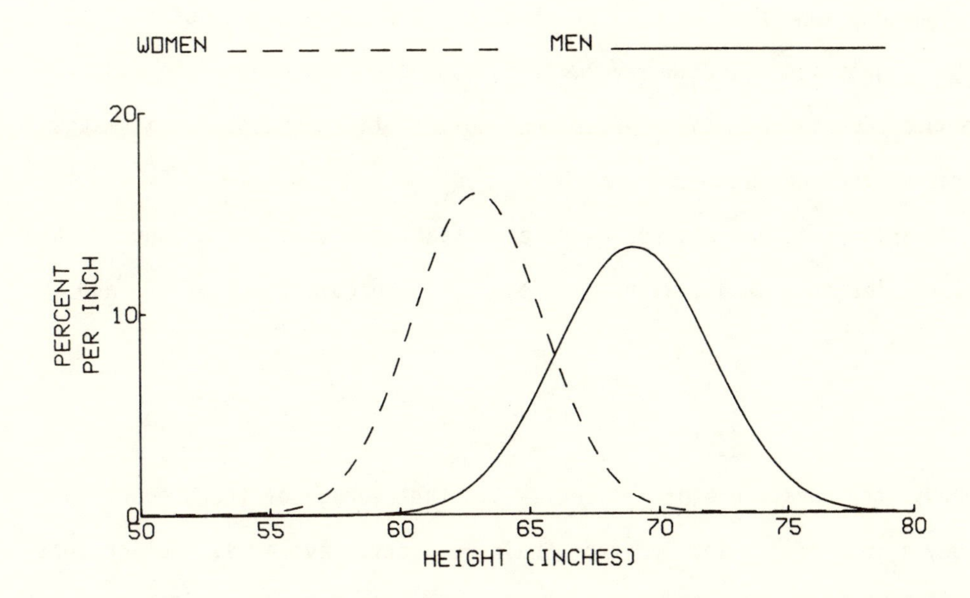

Exercise set 8

1. In one school, the children's heights followed the normal curve, with an average of 53 inches and a variance of 4 square inches.

   (a) μ = ____________ and σ = ____________

   (b) In original units, the curve is N(53,2)? or N(53,4)?

2. The N(25,16) curve has mean = ____________ and SD = ____________.

3. The N(16,25) curve has mean = ____________ and SD = ____________.

4. Show that the N(0,1) curve has the equation (1).

   For this reason, the standard normal curve is often denoted N(0,1).

9. Covariance

The next important objects of study are the correlation coefficient and the regression line. However, it is convenient to pause here and introduce <u>covariance</u>. This concept has no intuitive appeal, but it makes the mathematics go much more smoothly.

Covariance is defined for pairs of variables. Suppose x and y are two variables in a study, with n subjects. The covariance of x and y is defined as

$$\frac{1}{n} \sum_{i=1}^{n} (x_i - \bar{x})(y_i - \bar{y})$$

In words, the covariance of x and y is the average of the products of the deviations of x and y from their respective averages. Notice that covariance has funny units: the units for x times the units for y. Thus, the covariance for height and weight comes out in terms of inches $\times$ pounds.

As the name implies, the covariance says something about how x and y "covary", that is, vary together over the subjects in the study. If x and y are positively associated, then x tends to be above average when y is. So $x_i - \bar{x} > 0$ and $y_i - \bar{y} > 0$ tend to go together. Likewise, $x_i - \bar{x} < 0$ and $y_i - \bar{y} < 0$ tend to go together. Therefore, $(x_i - \bar{x})(y_i - \bar{y})$ tends to be positive. Thus, positive association between x and y is signalled by a positive covariance. Likewise, negative association is signalled by negative covariance.

The covariance between x and y is usually abbreviated as $\text{cov}(x,y)$. The formula for defining covariance is

(1) $$\operatorname{cov}(x,y) = \frac{1}{n}\sum_{i=1}^{n} (x_i - \bar{x})(y_i - \bar{y})$$

The next proposition gives an alternative way of computing the covariance. In words, the covariance is the average of the products, minus the product of the averages.

(2) Proposition. $\operatorname{cov}(x,y) = (\frac{1}{n}\sum_{i=1}^{n} x_i y_i) - (\bar{x}\bar{y})$

Proof. We expand in (1):

$$(x_i - \bar{x})(y_i - \bar{y}) = x_i y_i - \bar{x} y_i - x_i \bar{y} + \bar{x}\bar{y}$$

Now $\bar{x}$ and $\bar{y}$ are constants: they do not depend on the index of summation i. So proposition 5.7 on page 19 can be used:

$$\frac{1}{n}\sum_{i=1}^{n} (x_i-\bar{x})(y_i-\bar{y}) = (\frac{1}{n}\sum_{i=1}^{n} x_i y_i) - (\bar{x}\,\frac{1}{n}\sum_{i=1}^{n} y_i) - (\bar{y}\,\frac{1}{n}\sum_{i=1}^{n} x_i) + (\bar{x}\bar{y})$$

But

$$\frac{1}{n}\sum_{i=1}^{n} x_i = \bar{x} \quad \text{and} \quad \frac{1}{n}\sum_{i=1}^{n} y_i = \bar{y}$$

Substituting back,

$$\begin{aligned}\frac{1}{n}\sum_{i=1}^{n} (x_i - \bar{x})(y_i - \bar{y}) &= (\frac{1}{n}\sum_{i=1}^{n} x_i y_i) - \bar{x}\bar{y} - \bar{x}\bar{y} + \bar{x}\bar{y} \\ &= (\frac{1}{n}\sum_{i=1}^{n} x_i y_i) - (\bar{x}\bar{y})\end{aligned}$$

This completes the proof.

A useful fact about covariance can be stated as follows. Let u, v, and w be variables in a study. Let a and b be constants.

(3) Let $z_i = au_i + bv_i$ for all i. Then

$$\text{cov}(z,w) = a\ \text{cov}(u,w) + b\ \text{cov}(v,w)$$

To verify (3), notice that $\bar{z} = a\bar{u} + b\bar{v}$ by proposition 5.7 on page 19.

Now

$$z_i - \bar{z} = (au_i + bv_i) - (a\bar{u} + b\bar{v})$$
$$= a(u_i - \bar{u}) + b(v_i - \bar{v})$$

So

$$\text{cov}(z,w) = \frac{1}{n}\sum_{i=1}^{n} (z_i - \bar{z})(w_i - \bar{w})$$
$$= \frac{1}{n}\sum_{i=1}^{n} [a(u_i - \bar{u}) + b(v_i - \bar{v})](w_i - \bar{w})$$
$$= \frac{1}{n}\sum_{i=1}^{n} [a(u_i - \bar{u})(w_i - \bar{w}) + b(v_i - \bar{v})(w_i - \bar{w})]$$

Now use that proposition 5.7 again:

$$\text{cov}(z,w) = a[\frac{1}{n}\sum_{i=1}^{n} (u_i - \bar{u})(w_i - \bar{w})] + b[\frac{1}{n}\sum_{i=1}^{n} (v_i - \bar{v})(w_i - \bar{w})]$$
$$= a\ \text{cov}(u,w) + b\ \text{cov}(v,w)$$

This completes the argument.

Exercise set 9

1. Should the covariance between height and weight be positive or negative? Why?

2. A study is made relating fuel economy (highway m.p.g.) to automobile weight. Should the covariance be positive or negative? Why?

3. For the six subjects in Table 1 on page 22, compute: the variance of height, the variance of weight, the covariance of height and weight.

4. For a certain group of men, the following results are obtained: the average height is 68 inches, the average income is $14,000, and

$$\text{cov(height, income)} = 5{,}400 \text{ inch} \times \text{dollar}$$

(a) Is the association between height and income positive or negative?

(b) From the information given, can you determine the average of the products "height × income" for the men in the group? If so, what is it? If not, why not?

5. For another group of men, the following results are obtained: the average height is 69 inches, the average income is $17,000, and the average of the products "height × income" is 1,179,500 inch × dollar. From this information, can you determine the covariance of height and income? If so, what is it? If not, why not?

6. In a certain study, there are N subjects, and two variables u and v. The investigator defines t_u as follows:

$$t_u^2 = \frac{1}{N}\sum_{n=1}^{N} u_n^2 - \left(\frac{1}{N}\sum_{m=1}^{N} u_m\right)^2$$

Then t_u is the ________________ of the variable u. Fill in the blank, using one of the options below, and explain your choice.

(i) average

(ii) mean

[Continues on next page.]

(iii) mean square

(iv) root mean square

(v) standard deviation

(vi) variance

(vii) covariance

(viii) none of the foregoing

7. The formula for covariance is

$$\text{cov}(x,y) = \frac{1}{n} \sum_{i=1}^{n} (x_i - \bar{x})(y_i - \bar{y})$$

(a) Which part of this formula refers to the number of subjects in the study?

(b) Which part of this formula gives the average of x?

(c) Which part of this formula gives the deviation from average for subject i with respect to variable y?

(d) Which part of this formula gives the sum of the products of the deviations from average?

8. Prove that $\text{cov}(x,x) = \text{var } x$.

9. Prove that $\text{cov}(x,y) = \text{cov}(y,x)$.

10. If $x_i = c$ for all i, prove that $\text{cov}(x,y) = 0$.

11. If $z_i = mx_i + b$ for all i, prove that $\text{cov}(z,u) = m \text{ cov}(x,u)$.

12. Prove that $\text{cov}(x,y) = \frac{1}{n} \sum_{i=1}^{n} x_i(y_i - \bar{y}) = \frac{1}{n} \sum_{i=1}^{n} (x_i - \bar{x}) y_i$.

<u>Hint</u>: Imitate the proof of 7.4 on page 34.

13. Let u and v be two variables in a study. Let a, b and c be three constants. Define a new variable z as follows:

$$z_i = au_i + bv_i + c \quad \text{for all subjects } i$$

Prove: $\text{var } z = a^2 \text{ var } u + 2ab \text{ cov}(u,v) + b^2 \text{ var } v$.

14. In a certain study, there are N subjects; α and β are constants; ξ and ζ are variables. True or false, and explain:

(a) $\text{cov}(\xi,\zeta) = \frac{1}{N} \sum_{n=1}^{N} (\xi_n - \bar{\xi})(\zeta_n - \bar{\zeta})$

(b) $\text{cov}(\alpha\xi+\beta,\zeta) = \alpha \text{ cov}(\xi,\zeta) + \beta$

Pronunciation of greek letters: α is "alpha", β is "beta", ξ is "ksi", ζ is "zeta". For an explanation of $\alpha\xi + \beta$, see page 24.

15. True or false, and explain: in a study with m subjects,

(a) $\text{cov}(x,y) = \frac{1}{m} \sum_{n=1}^{m} (x_n - \bar{x})(y_n - \bar{y})$

(b) $\text{cov}(x,y) = \frac{1}{n} \sum_{i=1}^{n} (x_i - \bar{x})(y_i - \bar{y})$

10. The correlation coefficient

The object of this section is to express the correlation coefficient in sigma-notation, and establish some of its properties. The correlation coefficient r measures the strength of the linear relationship between two variables x and y in a study.[1] To compute r, the averages $\bar{x}$ and $\bar{y}$ are needed, as well as the standard deviations. Abbreviate

s_x for the standard deviation of x

s_y for the standard deviation of y.

Then, x and y must be converted to standard units: see exercise 6.9 on page 31 above. To convert x_i to standard units, just subtract the average $\bar{x}$ and divide by the standard deviation s_x, as follows:

$$\frac{x_i - \bar{x}}{s_x}$$

Likewise, y_i in standard units is

$$\frac{y_i - \bar{y}}{s_y}$$

Now r is the average product of the variables in standard units.

The correlation coefficient between x and y is written $r(x,y)$. The formula for r is:

$$(1) \qquad r(x,y) = \frac{1}{n} \sum_{i=1}^{n} \left(\frac{x_i - \bar{x}}{s_x}\right) \cdot \left(\frac{y_i - \bar{y}}{s_y}\right)$$

This formula can be rewritten in several equivalent ways. First, by

---

[1] Text, chapter 8.

formula 5.4 on page 18, with $1/s_x s_y$ for c and $(x_i - \bar{x}) \cdot (y_i - \bar{y})$ for x_i :

$$(2) \qquad r(x,y) = \frac{\frac{1}{n}\sum_{i=1}^{n} (x_i - \bar{x})\cdot(y_i - \bar{y})}{s_x s_y}$$

Next, by the definition 9.1 of covariance on page 43,

$$(3) \qquad r(x,y) = \frac{\operatorname{cov}(x,y)}{s_x s_y}$$

Finally, by formula 7.2 on page 33,

$$(4) \qquad r(x,y) = \frac{\operatorname{cov}(x,y)}{\sqrt{\operatorname{var} x}\,\sqrt{\operatorname{var} y}} .$$

We will now prove some useful facts about r .

(5) $r(x,y)$ is a pure number, without units.

This is because the units for x cancel out when we convert to standard units, and likewise for y .

(6) Interchanging x and y does not affect the correlation:

$$r(x,y) = r(y,x)$$

The reason: interchanging x and y just reverses the products in (1).

Next, we will show that multiplying all the values of x by the same constant does not change its correlation with y . Similarly, adding the same constant to all the values of x does not change its correlation with y . To state this as a formula, let x and y be two variables in a study. Let a, b, c and d be constants. Define new variables u and v as follows: for all subjects i ,

$$u_i = cx_i + d$$

$$v_i = ay_i + b$$

In more compact notation,

$$u = cx + d$$

$$v = ay + b$$

(7) Proposition. If c and a are positive constants, then

$$r(cx + d, ay + b) = r(x,y)$$

Proof. First, we will show that $u = cx + d$ is the same as x, in standard units. This is because $\bar{u} = c\bar{x} + d$ and $s_u = cs_x$, so

$$u_i - \bar{u} = (cx_i + d) - (c\bar{x} + d) = c(x_i - \bar{x})$$

Now

$$(8) \qquad \frac{u_i - \bar{u}}{s_u} = \frac{c(x_i - \bar{x})}{cs_x} = \frac{x_i - \bar{x}}{s_x}$$

(See exercise 6.10 on page 32.) The companion formula for $v = ay + b$ is

$$\frac{v_i - \bar{v}}{s_v} = \frac{y_i - \bar{y}}{s_y}$$

Combining this with (8) gives

$$\frac{1}{n}\sum_{i=1}^{n}\left(\frac{u_i - \bar{u}}{s_u}\right)\cdot\left(\frac{v_i - \bar{v}}{s_v}\right) = \frac{1}{n}\sum_{i=1}^{n}\left(\frac{x_i - \bar{x}}{s_x}\right)\cdot\left(\frac{y_i - \bar{y}}{s_y}\right)$$

The left side is $r(u,v)$; the right is $r(x,y)$. This completes the proof of (7).

The next fact is harder, but more interesting: correlations are always between -1 and 1 . The proof is postponed to section 14 .

(9) Proposition. $-1 \leq r \leq 1$.

We have glossed over one point: r is only defined when $s_x > 0$ and $s_y > 0$. If x is a constant and all the points lie on a vertical line, then $s_x = 0$ and $r(x,y)$ cannot be defined. Likewise, if y is a constant and all the points lie on a horizontal line, then $s_y = 0$ and $r(x,y)$ cannot be defined either. In this section, we have assumed $s_x > 0$ and $s_y > 0$.

Exercise set 10

1. Which of the following expressions, if any, equals $r(x,y)$? Give your reasons.

$$\text{(i)} \quad \frac{\frac{1}{n}\sum_{i=1}^{n} (x_i - \bar{x})(y_i - \bar{y})}{\sqrt{[(\frac{1}{n}\sum_{i=1}^{n} x_i^2) - \bar{x}^2] \cdot [\frac{1}{n}\sum_{i=1}^{n} (y_i - \bar{y})^2]}}$$

[Continues on next page.]

(ii) $$\frac{\sum_{i=1}^{n} x_i y_i - n\bar{x}\bar{y}}{\sqrt{[(\sum_{i=1}^{n} x_i^2) - (n\bar{x}^2)] \cdot [(\sum_{i=1}^{n} y_i^2) - (n\bar{y}^2)]}}$$

(iii) $$\frac{\sum_{i=1}^{n} x_i y_i - \bar{x}\bar{y}}{\sqrt{[(\sum_{i=1}^{n} x_i^2) - (\bar{x})^2] \cdot [(\sum_{i=1}^{n} y_i^2) - (\bar{y})^2]}}$$

2. For a certain group of 409 men, the following results are obtained:

   sum of heights = 28,359 inches

   sum of squares of heights = 1,969,716 square inches

   sum of weights = 64,938 pounds

   sum of squares of weights = 10,517,079 square pounds

   sum of (heights × weights) = 4,513,810 inch × pound.

   From this information, can you find the correlation between height and weight? If so, what is it? If not, why not?

3. For a certain group of 257 women, the correlation between height in inches and weight in pounds was 0.29.
   (a) What is the correlation between weight in pounds and height in inches? Or can this be obtained from the information given?
   (b) What is the correlation between weight in kilograms and height in centimeters? Or can this be obtained from the information given?

4. For two variables x and y in a certain study:

$$\sum_{i=1}^{n} x_i = 0 \qquad \sum_{i=1}^{n} x_i^2 = n$$

$$\sum_{i=1}^{n} y_i = 0 \qquad \sum_{i=1}^{n} y_i^2 = n$$

(a) $\bar{x} = ?$

(b) $\bar{y} = ?$

(c) $s_x = ?$

(d) $s_y = ?$

(e) True or false, and explain: In this study $r(x,y) = \frac{1}{n} \sum_{i=1}^{n} x_i y_i$.

5. A surveyor measures the lots in a certain sub-division. These lots are all rectangular. The data is analyzed by a statistician, as follows:

Average length = 107 ft, SD = 12 ft

Average width = 53 ft, SD = 7 ft

$r = 0.42$, $n = 89$

However, what the surveyor wants to know is the total area covered by the 89 lots. Can you help her?

6. Let x stand for height in the Health Examination Survey, measured in inches, and y for weight, measured in pounds. There were 6,672 subjects. True or false, and explain: the correlation between height and weight equals

$$\frac{\frac{1}{6,672} \sum_{i=1}^{6,672} (x_i - \bar{x}) y_i}{\sqrt{\frac{1}{6,672} \sum_{i=1}^{6,672} (x_i - \bar{x})^2} \sqrt{\left(\frac{1}{6,672} \sum_{i=1}^{6,672} y_i^2\right) - (\bar{y}^2)}}$$

7. In the formula of exercise 6, what if anything represents:

   (a) the height of the i-th subject

   (b) the deviation of subject i from average in height

   (c) the deviation of subject i from average in weight

   (d) the sum of the squares of the weights of the subjects

   (e) the sum of the squares of the heights of the subjects

   (f) the SD of the heights

   (g) the variance of the weights

8. Continuing exercises 6 and 7, the units for

   (a) the SD of weight are __________

   (b) the variance of height are __________

   (c) the covariance between height and weight are __________

   (d) the correlation between height and weight are __________

9. What happens to r if all the points lie on a horizontal line? A vertical line? Explain.

10. An investigator computes a correlation coefficient of -1.2 ; is anything wrong? Explain.

11. The regression line

Let x and y be two variables in a study. In many such studies, the average value of y will be linearly related to x . The regression line expresses this relationship in mathematical form.[1] The line can be used to help summarize existing data, and to predict y from x for new subjects. An example is the relationship between scores on the Law School Aptitude Test (LSAT), and first-year scores in a law school. Let

x_i = LSAT score for student i

y_i = first-year score for student i

The regression line can be fitted to data on a sample group of students, as we will discuss in this section. The line can then be used to predict first-year scores from LSAT scores for future applicants.[2]

The regression line for predicting y from x has the equation[3]

(1) predicted $y = mx + b$, where $m = r \cdot s_y/s_x$ and $b = \bar{y} - m\bar{x}$

Throughout this section,

(2) m = slope of regression line for predicting y from x $= r \cdot s_y/s_x$

(3) b = intercept of regression line for predicting y from x $= \bar{y} - m\bar{x}$

---

[1]See chapters 10 and 12 in the text.

[2]This very brief discussion cannot do justice to the complications which occur in practice.

[3]See chapter 12 of the text.

Equation (1) can be reorganized a bit:

$$(4) \qquad y = \bar{y} + m(x-\bar{x})$$

Substituting $x = \bar{x}$, we get $y = \bar{y}$: the regression line goes through the point of averages $(\bar{x}, \bar{y})$.

Another formula for m is often useful:

$$(5) \qquad m = \text{cov}(x,y)/\text{var}\, x$$

To verify this, we start from formula 10.4 for r on page 49:

$$r = \frac{\text{cov}(x,y)}{\sqrt{\text{var } x}\,\sqrt{\text{var } y}}$$

But $s_y = \sqrt{\text{var } y}$ and $s_x = \sqrt{\text{var } x}$, by formula 7.2 on page 33. Substitution into (2) gives (5).

The value of y predicted by equation (1) from x_i is usually denoted by a hat:

$$(6) \qquad \hat{y}_i = mx_i + (\bar{y} - m\bar{x}) = \bar{y} + m(x_i - \bar{x})$$

The residual, or prediction error, is usually denoted by e_i :

$$(7) \qquad e_i = \text{actual} - \text{predicted} = y_i - \hat{y}_i$$

To sum up: subject i has an actual y-value of y_i, and a predicted value--from the regression equation--of $\hat{y}_i$. For subject i, the regression method makes an error of $e_i = y_i - \hat{y}_i$. (In ordinary usage, the error is "predicted - actual". However, the present definition of e_i makes the mathematics smoother.)

There is one technical point we have glossed over. If x is constant, so $\text{var } x = 0$, the regression line for predicting y from x is not defined. In this section, we assume $\text{var } x > 0$.

Figure 1. A prediction error is a vertical distance.

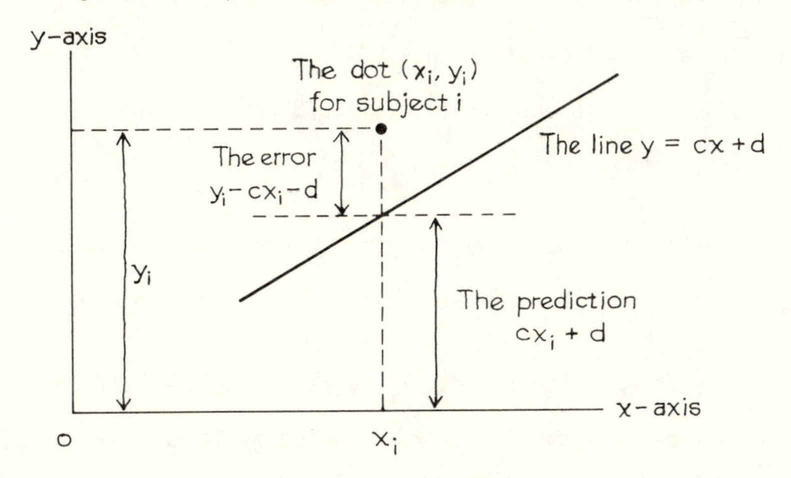

What is special about the regression line? In principle, y could be predicted from x using any line, for instance,

$$\text{predicted } y = cx + d$$

As figure 1 shows, the prediction error for subject i will then be

$$y_i - cx_i - d$$

The average of the squared errors will be

$$\text{MSE} = \frac{1}{n}\sum_{i=1}^{n} (y_i - cx_i - d)^2$$

MSE is an abbreviation for Mean Squared Error.

For which line is the MSE smallest? The answer is, the regression line. This remarkable fact was first proved by Carl Friedrich Gauss (Germany, 1777-1855). The argument is long, and it is postponed to section 13.

Exercise set 11

1. For a certain group of 409 men, the following results are obtained:

   sum of heights = 28,359 inches

   sum of squares of heights = 1,969,716 square inches

   sum of weights = 64,938 pounds

   sum of squares of weights = 10,517,079 square pounds

   sum of (heights × weights) = 4,513,810 inch × pound

   From this information, can you find the regression line for predicting weight from height? If so, what is it? If not, why not?

2. For the study in exercise 1, let x represent height, and y weight. An investigator wants the line $cx + d$ which minimizes the <u>sum of squares for error</u> (SSE):

$$\text{SSE} = \sum_{i=1}^{409} (y_i - cx_i - d)^2$$

   Can you help her? Explain carefully.

3. Hooke's law states that, when a load (weight) is placed on a spring,

   length under load = constant · load + length under no load

   The following experimental results are obtained:

Table 1. Data on Hooke's law

Load (kg)	Length (cm)
0	287.12
1	287.18
1	287.16
3	287.25
4	287.33
4	287.35
6	287.40
12	287.75

(a) The two lengths for a load of 1 kg differ. Why might that be?

(b) Find the regression equation for predicting length from load.

(c) Use the equation to predict length at the following loads: 2 kg, 3 kg, 5 kg, 105 kg.

(d) For a load of 3 kg, the answer to (b) is different from the number in the table. Under the load of 3 kg, would you use the number in the table, or the regression equation? Explain carefully.

(e) Estimate the length of the spring under no load.

(f) Estimate the constant in Hooke's law.

Note: To answer parts (a) and (d), you may have to refer to section 2 of chapter 12 in the text.

4. For a certain study with n subjects:

$$\sum_{i=1}^{n} x_i = 0 \qquad \sum_{i=1}^{n} x_i^2 = n$$

$$\sum_{i=1}^{n} y_i = 0 \qquad \sum_{i=1}^{n} y_i^2 = n$$

[Continues on next page.]

True or false, and explain: in this study, the equation of the regression line is $y = rx$.

5. Show that the slope of the regression line can be expressed as

$$m = \frac{1}{n}\sum_{i=1}^{n} (x_i - \bar{x})\, y_i \,/ s_x^2 .$$

6. For a certain group of people, the regression line for predicting income (dollars) from education (year of schooling completed) is $y = mx + b$. The units for b are ________. The units for m are ________.

7. The Health Examination Survey had 3,091 male subjects. Let x stand for height and y for systolic blood pressure. The regression line for predicting blood pressure from height for the men has the equation $y = mx + b$, where

$$m = \frac{\frac{1}{3,091}\sum_{i=1}^{3,091} (x_i - \bar{x})(y_i - \bar{y})}{\frac{1}{3,091}\sum_{i=1}^{3,091} (x_i - \bar{x})^2}$$

What part of this formula, if any, represents

(a) the variance of height?

(b) the variance of blood pressure?

(c) the covariance between height and blood pressure?

(d) the covariance between height and weight?

8. Which of the following gives the slope of the regression line for predicting y from x ? Explain your choices.

(a) $$\frac{\sum_{i=1}^{n} (x_i - \bar{x})(y_i - \bar{y})}{\sum_{i=1}^{n} (x_i - \bar{x})^2}$$

[Continues on next page.]

(b) $\dfrac{\sum_{i=1}^{n} (x_i - \bar{x})y_i}{\sum_{i=1}^{n} (x_i - \bar{x})^2}$

(c) $\dfrac{\sum_{i=1}^{n} x_i(y_i - \bar{y})}{\sum_{i=1}^{n} (x_i - \bar{x})^2}$

(d) $\dfrac{\sum_{i=1}^{n} x_i y_i}{\sum_{i=1}^{n} (x_i - \bar{x})^2}$

(e) $\dfrac{\sum_{i=1}^{n} (x_i - \bar{x})(y_i - \bar{y})}{\sum_{i=1}^{n} x_i^2}$

(f) $\dfrac{\sum_{i=1}^{n} x_i y_i - \bar{x}\bar{y}}{\sum_{i=1}^{n} (x_i - \bar{x})^2}$

(g) $\dfrac{\sum_{i=1}^{n} x_i y_i - n\bar{x}\bar{y}}{\sum_{i=1}^{n} (x_i - \bar{x})^2}$

9. True or false, and explain: among all lines for predicting y from x, the regression line has the smallest r.m.s. error. (Hard.)

10. True or false, and explain: among all lines for predicting y from x, the regression line has the smallest <u>average absolute error</u>. (Hard.)

    Definition: If the line $cx + d$ is used to make the prediction, the average absolute error is

$$\frac{1}{n} \sum_{i=1}^{n} |y_i - cx_i - d|$$

11. Let x and y be two variables in a study. One investigator is interested in predicting y from x, and computes a regression line for this purpose. Another investigator is interested in predicting x from y, and computes a regression line for that purpose. Are the two lines the same? If so, why? If not, compute the ratio of the two slopes.

12. The normal equations

For the reader who knows calculus, here is an argument to prove that among all lines, the regression line minimizes the mean squared error: an algebraic proof will be given in section 13. We assume $\text{var } x > 0$. The problem is to find the c and d which minimize

$$(1) \qquad \frac{1}{n} \sum_{i=1}^{n} (y_i - cx_i - d)^2$$

There is something a bit tricky here: for present purposes, x_i and y_i must be treated as constants, because they depend on the data and are therefore beyond our control. In searching for the minimum, all we can vary is the c and d in the formula for the line. Therefore, we must differentiate the expression (1) with respect to c and d, and set both derivatives equal to zero. Differentiating (1) with respect to c gives

$$(2) \qquad -2 \frac{1}{n} \sum_{i=1}^{n} x_i(y_i - cx_i - d)$$

Setting this equal to zero gives the equation

$$(3) \qquad \frac{1}{n} \sum_{i=1}^{n} x_i(y_i - cx_i - d) = 0$$

Differentiating (1) with respect to d gives

$$(4) \qquad -2 \frac{1}{n} \sum_{i=1}^{n} (y_i - cx_i - d)$$

Setting this equal to zero gives the equation

$$(5) \qquad \frac{1}{n} \sum_{i=1}^{n} (y_i - cx_i - d) = 0$$

Equations (3) and (5) are called the normal equations; however, they have nothing to do with the normal curve. Statisticians like to call things "normal"; they find it reassuring.

Equation (5) can be simplified, using proposition 5.7 on page 19:

$$\bar{y} - c\bar{x} - d = 0$$

Solving for d,

$$d = \bar{y} - c\bar{x} \qquad (6)$$

Substitute this into (3):

$$\begin{aligned} y_i - cx_i - d &= y_i - cx_i - (\bar{y} - c\bar{x}) \\ &= (y_i - \bar{y}) - c(x_i - \bar{x}) \end{aligned}$$

Now we use proposition 5.7 to simplify (3):

$$\frac{1}{n}\sum_{i=1}^{n} x_i(y_i - \bar{y}) - c\,\frac{1}{n}\sum_{i=1}^{n} x_i(x_i - \bar{x}) = 0 \qquad (7)$$

Rearrange this:

$$c\,\frac{1}{n}\sum_{i=1}^{n} x_i(x_i - \bar{x}) = \frac{1}{n}\sum_{i=1}^{n} x_i(y_i - \bar{y}) \qquad (8)$$

But on the left side, the coefficient of c is var x, by formula 7.4 on page 34 above. The expression on the right side of (8) is $\text{cov}(x,y)$, by exercise 9.12 on page 46. So (8) comes down to

$$c \cdot \text{var}\, x = \text{cov}(x,y) \qquad (9)$$

Thus, $c = \text{cov}(x,y)/\text{var}\, x$. And this is the slope m of the regression line, by formula 11.5 on page 56. Finally, go back to (6):

$$d = \bar{y} - c\bar{x} = \bar{y} - m\bar{x}$$

Thus, d is the intercept of the regression line, by formula 11.3 on page 55.

Exercise set 12

1. Use calculus to find the d which minimizes

$$\frac{1}{n}\sum_{i=1}^{n}(x_i - d)^2$$

What is d called?

2. Use calculus to find the c which minimizes

$$\frac{1}{n}\sum_{i=1}^{n}(y_i - cx_i)^2$$

Is c the slope of the regression line for predicting y from x? Explain.

Technical note. The argument in this section identifies a critical point for the function (1). Why is it a minimum, rather than a maximum or a saddle point? To answer this by the techniques of calculus, it is necessary to look at the matrix of second-order derivatives; from (2) and (4), this is 2M, where

$$M = \begin{pmatrix} \frac{1}{n}\sum_{i=1}^{n} x_i^2 & \frac{1}{n}\sum_{i=1}^{n} x_i \\ \frac{1}{n}\sum_{i=1}^{n} x_i & 1 \end{pmatrix}$$

Then, it can be shown that M is strictly positive definite.

Many advanced calculus texts will have a statement of this criterion for an extreme point to be a minimum. However, a careful proof comes down to completing the square, as in the next section.

13. More on regression

This section is somewhat technical; readers can skip to section 14. The main objective is to prove that among all lines for predicting y from x, the regression line has the smallest MSE.

The prediction errors e_i for the regression line were defined in formula 11.7 on page 56:

$$(1) \qquad e_i = \text{actual} - \text{predicted} = y_i - \hat{y}_i$$

We have to study these e_i quite carefully. Remember formula 11.6 on page 56:

$$(2) \qquad \hat{y}_i = \bar{y} + m(x_i - \bar{x})$$

Substitute (2) into (1):

$$(3) \qquad e_i = (y_i - \bar{y}) - m(x_i - \bar{x})$$

Next, the average error is 0:

$$(4) \qquad \bar{e} = \frac{1}{n} \sum_{i=1}^{n} e_i = 0$$

To see this, use proposition 5.7 on page 19 above, applied to (3):

$$\frac{1}{n} \sum_{i=1}^{n} e_i = \frac{1}{n} \sum_{i=1}^{n} (y_i - \bar{y}) - m \frac{1}{n} \sum_{i=1}^{n} (x_i - \bar{x})$$

In this application, we substitute:

e_i for z_i, $y_i - \bar{y}$ for y_i, $x_i - \bar{x}$ for x_i

1 for b, $-m$ for c, 0 for d

But the average deviation from average is 0, by proposition 5.8:

$$\frac{1}{n}\sum_{i=1}^{n}(y_i - \bar{y}) = 0 \quad \text{and} \quad \frac{1}{n}\sum_{i=1}^{n}(x_i - \bar{x}) = 0$$

This proves (4).

Next, the prediction errors e_i do not have any linear relationship with the data x_i. Mathematically, the correlation between e and x is zero, because $\text{cov}(e,x) = 0$. The proof is a bit intricate, and is broken up into several smaller steps, called "lemmas".

(5) Lemma. $\frac{1}{n}\sum_{i=1}^{n}(x_i - \bar{x})e_i = 0$

Proof. Start from (3), and use proposition 5.7 on page 19:

$$(6) \quad \frac{1}{n}\sum_{i=1}^{n}(x_i-\bar{x})e_i = \frac{1}{n}\sum_{i=1}^{n}(x_i-\bar{x})(y_i-\bar{y}) - m\frac{1}{n}\sum_{i=1}^{n}(x_i-\bar{x})^2$$

The first term on the right is $\text{cov}(x,y)$, by definition 9.1 on page 43. The second is $m \cdot \text{var}\ x$, by definition 7.1 on page 33. And $m = \text{cov}(x,y)/\text{var}\ x$ by 11.5 on page 56. Substituting back into (6),

$$\begin{aligned}\frac{1}{n}\sum_{i=1}^{n}(x_i - \bar{x})e_i &= \text{cov}(x,y) - \frac{\text{cov}(x,y)}{\text{var}\ x}\text{var}\ x \\ &= \text{cov}(x,y) - \text{cov}(x,y) \\ &= 0\end{aligned}$$

This completes the proof.

(7) Lemma. $\text{cov}(x,e) = \text{cov}(e,x) = 0$

Proof. First, $\text{cov}(x,e) = 0$ by (5) and exercise 9.12 on page 46. Next, $\text{cov}(x,e) = \text{cov}(e,x)$ by exercise 9.9, completing the proof.

(8) Proposition. $r(x,e) = 0$

Proof. Substitute (7) into formula 10.3 for r on page 49 above. This completes the proof.

We are now going to compare the line $cx + d$ to the regression line. Let f_i be the prediction error made by the line $cx + d$ at subject i :

(9) $$f_i = y_i - cx_i - d$$

In view of proposition 5.7 on page 19,

(10) $$\bar{f} = \bar{y} - c\bar{x} - d$$

Remember that e_i is the prediction error made by the regression line. Our next job is to see how f_i is related to e_i .

(11) <u>Lemma</u>. $f_i = e_i + (m-c)\cdot x_i + (\bar{y} - m\bar{x} - d)$

<u>Proof</u>. This is a nasty bit of algebra:

$$\begin{aligned} f_i &= y_i - cx_i - d \\ &= [(y_i-\bar{y}) - m(x_i-\bar{x})] + (m-c)x_i + (\bar{y} - m\bar{x} - d) \\ &= e_i + (m-c)\cdot x_i + (\bar{y} - m\bar{x} - d) \qquad \text{by (3)} \end{aligned}$$

This completes the proof.

(12) <u>Lemma</u>. $\text{var } f = \text{var } e + (m-c)^2\cdot\text{var } x$

<u>Proof</u>. First, $\bar{y} - m\bar{x} - d$ is a constant. So, applying exercise 9.13 on page 47 to (11),

$$\text{var } f = \text{var } e + 2(m-c)\cdot\text{cov}(e,x) + (m-c)^2\cdot\text{var } x$$

But $\text{cov}(e,x) = 0$ by (7), completing the proof.

(13) <u>Lemma</u>. $\text{var } f = (\frac{1}{n}\sum_{i=1}^{n} f_i^2) - (\bar{y} - c\bar{x} - d)^2$

<u>Proof</u>. Use (10), and formula 7.3 on page 34.

(14) Lemma. $\text{var } e = \frac{1}{n}\sum_{i=1}^{n} e_i^2$

Proof. Use (4), and formula 7.3 on page 34.

(15) Lemma. $\frac{1}{n}\sum_{i=1}^{n} f_i^2 = \frac{1}{n}\sum_{i=1}^{n} e_i^2 + (m-c)^2 \cdot \text{var } x + (\bar{y} - c\bar{x} - d)^2$

Proof. Substitute (13) and (14) into (12); then bring $(\bar{y} - c\bar{x} - d)^2$ to the other side.

The worst part of the argument is over. Remember that e_i represents the prediction error made by the regression line for subject i, and f_i is the error made by the competitor line $cx + d$. So $\frac{1}{n}\sum_{i=1}^{n} e_i^2$ is the MSE of the regression line; and $\frac{1}{n}\sum_{i=1}^{n} f_i^2$ is the MSE of the competitor line $cx + d$. Thus, (15) can be rewritten as follows:

(16) MSE of competitor line $cx + d$

$$= \text{MSE of regression line} + (m-c)^2 \cdot \text{var } x + (\bar{y} - c\bar{x} - d)^2$$

We can now prove Gauss's theorem.

(17) Proposition. Among all lines for predicting y from x, the regression line has the smallest mean square error.

Proof. Take any other line for predicting y from x, like $cx + d$. We have to prove that its MSE is bigger than the MSE of the regression line. Equation (16) makes duck soup out of this problem. Indeed, squares and variances can never be negative, so the extra two terms on the right side of (16) can never be negative. In other words, the line $cx + d$ can never have a smaller MSE than the regression line.

Even ties are impossible. For suppose the extra terms are 0 . First,

$$(m-c)^2 \cdot \text{var } x = 0$$

$$(m-c)^2 = 0$$

$$c = m$$

$$c = \text{the slope of the regression line}$$

Then

$$(\bar{y} - c\bar{x} - d)^2 = 0$$

$$\bar{y} - c\bar{x} - d = 0$$

$$d = \bar{y} - c\bar{x}$$

$$d = \bar{y} - m\bar{x} \quad \text{because} \quad c = m$$

$$d = \text{the intercept of the regression line}$$

That is to say, if the line $cx + d$ has the same MSE as the regression line, then it is the regression line.

[There are no exercises to this section.]

14. The analysis of variance for regression

The object in this section is to derive a formula for the r.m.s. prediction error made by the regression line.[1] To have a definite example, consider the regression of weight on height for the 411 men aged 18 to 24 in the Health Examination Survey of 1960. The summary statistics:

average height = 68 inches SD = 2.5 inches

average weight = 158 pounds SD = 25 pounds

$r = 0.36$

The equation of the regression line:

$$\text{predicted weight} = (3.6 \text{ lbs/in}) \times \text{height} - 86.8 \text{ lbs}$$

Algebraically,

$$\hat{y}_i = mx_i + b \tag{1}$$

where

m = slope of regression line = 3.6 lbs/in

b = intercept of regression line = -86.8 lbs

x_i = height of subject i

$\hat{y}_i$ = predicted weight of subject i

For later use, let

y_i = actual weight of subject i.

[1]See page 171 of the text.

The error or residual of the regression line at subject i is

$$(2) \qquad e_i = y_i - \hat{y}_i$$

So

$$(3) \qquad y_i = \hat{y}_i + e_i$$

In words,

actual weight = predicted weight + residual .

As the formula shows, there are two sources of variation in the weights:

- variation along the regression line, that is, in the predicted weights $\hat{y}$;
- variation around the regression line, that is, in the residuals e.

These two sources of variation are additive, in the following technical sense:

$$(4) \qquad \text{var } y = \text{var } \hat{y} + \text{var } e$$

This formula is called an analysis of variance: the variance of weight is decomposed--analyzed--into two terms: the variance of predicted weights, and the variance of the residuals.

Why is the formula (4) true? The starting point in a proof is 13.7 on page 67:

$$(5) \qquad \text{cov}(x,e) = 0$$

Remember formula (1):

$$(6) \qquad \hat{y}_i = mx_i + b$$

Using exercise 9.11 on page 46, we can combine (5) and (6):

$$\text{(7)} \qquad \text{cov}(\hat{y},e) = 0$$

By formula (3) above, $y_i = \hat{y}_i + e_i$. Using exercise 9.13 on page 47,

$$\begin{aligned} \text{var } y &= \text{var } \hat{y} + \text{var } e + 2\ \text{cov}(\hat{y},e) \\ &= \text{var } \hat{y} + \text{var } e \end{aligned}$$

This completes the proof of (4).

Variance is a technical idea, rather than a measure of spread. However, formula (4) does say something very useful about spread. To see what this is takes some detective work. Formula (1), and exercise 7.7 on page 36, show

$$\text{(8)} \qquad \text{var } \hat{y} = m^2 \text{ var } x$$

Use formula 11.2 on page 55 above:

$$\begin{aligned} m^2 &= r^2(s_y)^2/(s_x)^2 \\ &= r^2 \cdot (\text{var } y)/(\text{var } x) \end{aligned}$$

Substituting back into (8),

$$\text{(9)} \qquad \text{var } \hat{y} = r^2 \cdot \text{var } y$$

By formula (4),

$$\text{var } y = \text{var } \hat{y} + \text{var } e$$

So

$$\text{var } e = \text{var } y - \text{var } \hat{y}$$
$$= \text{var } y - r^2 \cdot \text{var } y$$

That is,

(10) $$\text{var } e = (1 - r^2) \cdot \text{var } y$$

Taking square roots,

(11) the standard deviation of e equals $\sqrt{1 - r^2}$ times the standard deviation of y.

Let us now back up and see what the standard deviation of e represents. As formula 13.4 on page 66 shows, e has an average of 0. So the standard deviation of e equals $\sqrt{\frac{1}{n} \sum_{i=1}^{n} e_i^2}$. But remember the definition (2): namely, e_i represents the prediction error made by the regression line for subject i. In other words, the standard deviation of e equals the r.m.s. of the prediction errors made by the regression line. So formula (11) can be restated as follows:

(12) The root-mean-square error made by the regression line for predicting y from x is $\sqrt{1 - r^2}$ times the standard deviation of y.

We have derived this important fact from (4).

A by-product of the argument is interesting:

(13) <u>Corollary</u>. $-1 \leqq r \leqq 1$

<u>Proof</u>. In (10), var y is positive and var e cannot be negative, so

$$1 - r^2 \geqq 0$$
$$r^2 \leqq 1$$
$$-1 \leqq r \leqq 1$$

This completes the proof.

Equation (4) is represented graphically in figure 1 below. The hypotenuse represents s_y, the standard deviation of y. The bottom edge of the triangle represents the standard deviation of $\hat{y}$, which equals $r \cdot s_y$, by taking square roots in equation (9). The right edge of the triangle represents the standard deviation of e, which equals $\sqrt{1 - r^2} \cdot s_y$, by (11). The Theorem of Pythagoras $(c^2 = a^2 + b^2)$ is the geometric counterpart to equation (4). In these terms, equation (4) is

$$\text{(14)} \qquad (\text{SD of } y)^2 = (\text{SD of } \hat{y})^2 + (\text{SD of } e)^2$$

More about this in section 16.

Figure 1. An analysis of variance.

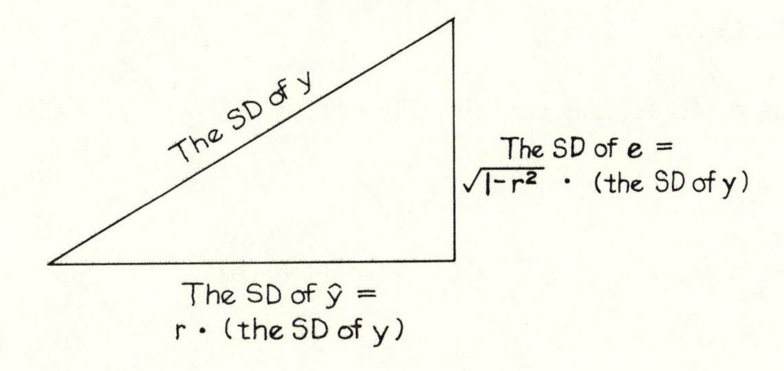

We will now apply these concepts to the example given at the beginning of the section, namely a regression of weight on height. The total variance of weight is

$$\text{var } y = (s_y)^2 = (25 \text{ lb})^2 = 625 \text{ lb}^2$$

The variance of the predicted weights is

$$\begin{aligned} \text{var } \hat{y} &= r^2 \cdot \text{var } y & \text{by (9)} \\ &= (0.36)^2 \times 625 \text{ lb}^2 \\ &= 81 \text{ lb}^2 \end{aligned}$$

The variance of the residuals is

$$\begin{aligned} \text{var } e &= (1 - r^2) \times \text{var } y & \text{by (10)} \\ &= [1 - (0.36)^2] \times 625 \text{ lb}^2 \\ &= 544 \text{ lb}^2 \end{aligned}$$

Note that $81 + 544 = 625 \text{ lb}^2$, the total variance of weight.

The SD of the residuals, and hence the r.m.s. error of the regression line for predicting weight from height, is $\sqrt{544} \approx 23$ lbs. In other words, if you were to use the regression line to predict weight from height, you should expect to make errors on the order of 23 lbs. That is a lot, but then the correlation of 0.36 is quite low.[1]

Exercise set 14

1. True of false, and explain: (SD of y) = (SD of $\hat{y}$) + (SD of e) .

[1]For more discussion, see Chapter 11 of the text.

2. A statistical analysis is made on SAT scores for a group of 100 entering freshmen; let x be the math SAT score, and y the verbal SAT score. The following results were obtained:

$$\sum_{i=1}^{100} x_i = 64{,}300 \qquad \sum_{i=1}^{100} y_i = 59{,}800$$

$$\sum_{i=1}^{100} x_i^2 = 42{,}341{,}000 \qquad \sum_{i=1}^{100} y_i^2 = 36{,}570{,}800$$

$$\sum_{i=1}^{100} x_i y_i = 39{,}171{,}400$$

A regression analysis is made to predict the verbal SAT from the math SAT.

(a) What is the total variance of the VSAT scores?

(b) What is the variance of the predicted VSAT scores?

(c) What is the variance of the residuals?

(d) What is the r.m.s. error of the regression?

3. Prove that

$$\frac{1}{n}\sum_{i=1}^{n} e_i^2 = \frac{1}{n}\sum_{i=1}^{n} (y_i - \bar{y})^2 - s_x^2\, m^2$$

Hint: Use (8) and (10).

15. Explained variance

The object of this section is to discuss the concept of "explained variance". The starting point is formula 14.4 on page 72:

$$(1) \qquad \text{var } y = \text{var } \hat{y} + \text{var } e$$

Here, the regression line is used to predict y from x; the predicted values are denoted by $\hat{y}$, and the prediction errors by e. In this formula,

> var y is called the <u>total variance</u>
>
> var $\hat{y}$ is called the <u>explained variance</u>
>
> var e is called the <u>unexplained</u> (or <u>residual</u>) <u>variance</u>.

That is, var $\hat{y}$ is that part of the variance attributable to variation along the regression line--changes in x: indeed, formula 14.1 shows that $\hat{y}$ can change only if x changes. By formula 14.9,

$$(2) \qquad \text{var } \hat{y} = r^2 \cdot \text{var } y$$

So the following terminology is often used:

> (3) r^2 is the fraction of variance that is "explained" in the regression.

Investigators who use this terminology also tend to use r^2 as a measure of linear association.

The image of "explained variance" is used very frequently, so it is important to remember the limitations of this idea:

- r^2 does not say anything about the absolute level of the variables.

An example is the Skeels-Skodak study on IQ's of adopted children.[1] The effect of the environment provided by the foster parents was to add ten or twenty points to the IQ's of the adopted children. However, in a regression of these IQ's on environmental variables (like the educational levels of the foster parents), almost none of the variance would be "explained": r^2 would be close to zero. The reason is that the effect of the environment was almost constant across children--it contributed almost no extra variation.

- "Explained" is used only in a technical sense.

For example, consider a regression of the area of some rectangles on their perimeter.[2] If r turns out to be 0.7, say, then $r^2 \approx 0.50$, and 50% of the variance in area is "explained" by perimeter. In fact, nothing at all has been explained by this regression. Half the variation may be along the regression line, but that particular regression line was silly.

- Variance is not a good way to measure variation, so r^2 is not a good descriptive statistic.

Figure 1 on the next page shows why. In the scatter plot at the top left, the correlation coefficient is 0.95. Now look at the three histograms. The top one is for the actual y-values. Its spread represents the total

[1]See pages 139-141 of the text.
[2]See pages 195-196 of the text.

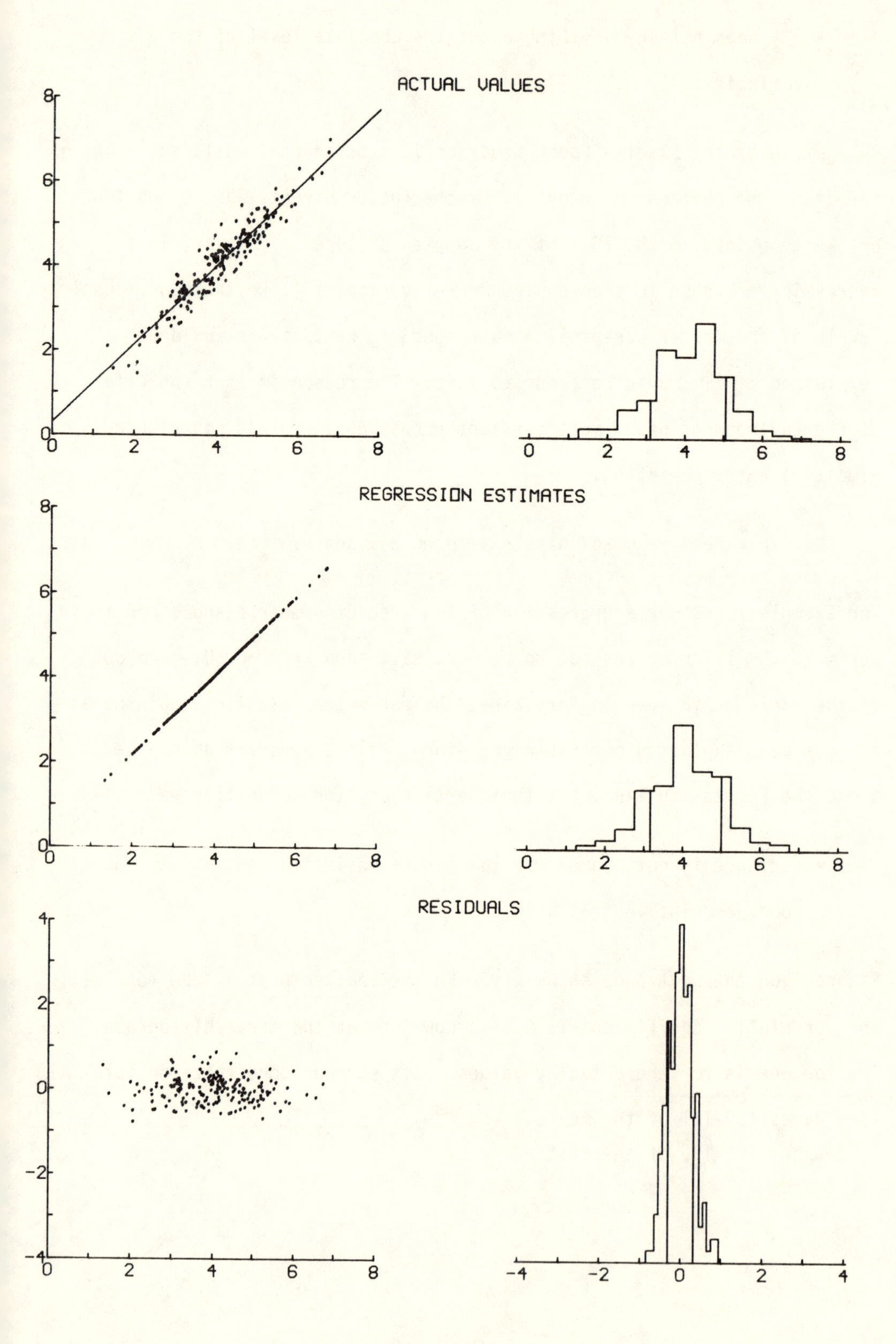
ACTUAL VALUES
REGRESSION ESTIMATES
RESIDUALS

variation. The middle one is for the predicted $\hat{y}$-values. The bottom one is for the residuals e. The variance of the residuals is $1 - r^2 \approx 10\%$ of the total. However, the SD of the residuals is $\sqrt{1 - r^2} \approx 30\%$ of the total. The spread in the bottom histogram is about 30% of the spread in the top--not 10%. (The region within one SD of average is marked by vertical lines.)

This discussion points to the basic problem with using r^2 as a measure of linear association: r^2 compares variances, and variance is not a good measure of spread. For example, in 1979 the average weight of American men aged 18 to 24 was about 170 pounds, with an SD of 30 pounds.[1] So far so good: a typical person in this population weighs something like 170 pounds, give or take 30 pounds or so. But the variance of weight in this population is--900 square pounds. Neither this magnitude nor its units are good at indicating spread.

Exercise set 15

1. True or false, and explain:

   (a) the variance of y equals the explained variance plus the MSE of the regression line.

   (b) the SD of y equals the SD of $\hat{y}$ plus the r.m.s. error of the regression line.

2. If $r = 0.7$, the percentage of variance explained equals ________ %; furthermore, the r.m.s. error of the regression line equals ______ % of the SD of y.

---

[1] Data from HANES--the Health and Nutrition Examination Survey, Vital and Health Statistics, Series 11, Number 208, 1979.

3. If $r = 0.99$, the percentage of variance explained equals ______ %; furthermore, the r.m.s. error of the regression line equals ______ % of the SD of y.

4. In the example of section 14, a regression line is used to predict weight from height. The percentage of variance explained in this regression is __________.

5. Continuing exercise 4, suppose another regression line is fitted to the same data, to predict height from weight. The percentage of variance explained in this regression is __________.

16. The geometry of regression

Though technical, this section may prove quite useful. Suppose that x and y are two variables in a study with n subjects. A scatter diagram can be used to represent the data, by means of n points in a two-dimensional graph. In this section, however, we wish to represent the same data by means of two points in n-dimensional space. First, a brief introduction to n-dimensional space.

The plane is two-dimensional: a point is represented by two coordinates (x_1, x_2), as in figure 1a. These coordinates measure distance from the origin along the two coordinate axes. Ordinary space is three-dimensional: a point is represented by three coordinates (x_1, x_2, x_3), as in figure 1b: there are three coordinate axes. In n-dimensional space, a point is represented by n coordinates $(x_1, x_2, \ldots, x_n)$. There are n coordinate axes. This space is a mathematical fiction, but a useful one.

Figure 1. The plane and ordinary space.

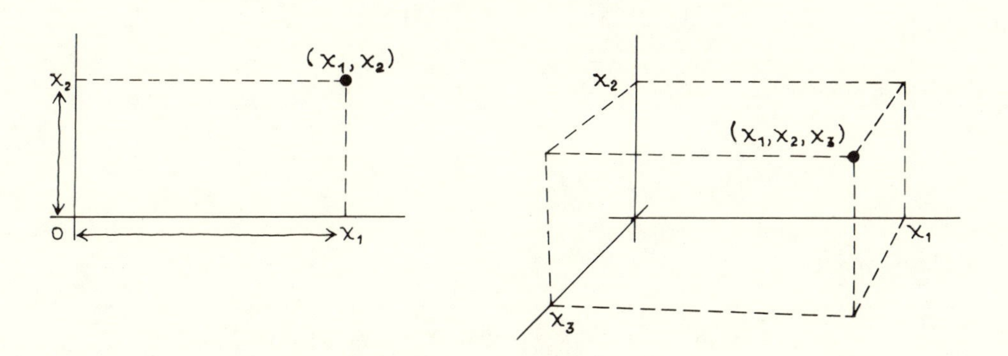

Suppose x and y are points in n-dimensional space. What is the distance between x and y? In two dimensions,

$$x = (x_1, x_2) \quad \text{and} \quad y = (y_1, y_2)$$

The distance is

$$\sqrt{(y_1 - x_1)^2 + (y_2 - x_2)^2}$$

This is the algebraic counterpart of the Theorem of Pythagoras: see figure 2.

Figure 2. Distance in two-dimensional space.

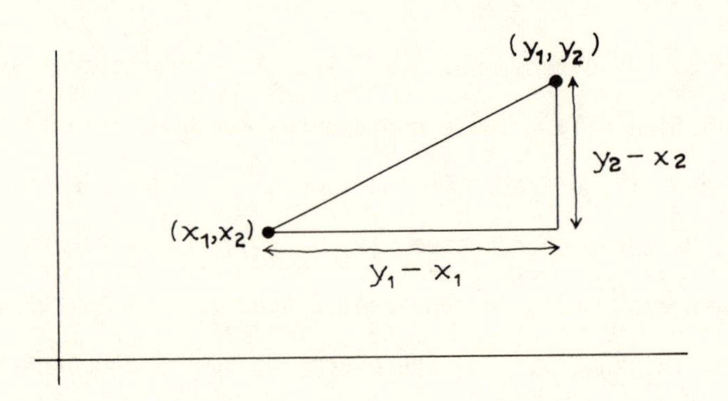

In three dimensions,

$$x = (x_1, x_2, x_3) \quad \text{and} \quad y = (y_1, y_2, y_3)$$

The distance between x and y is

$$\sqrt{(y_1 - x_1)^2 + (y_2 - x_2)^2 + (y_3 - x_3)^2}$$

In n dimensions,

$$x = (x_1, x_2, \ldots, x_n) \quad \text{and} \quad y = (y_1, y_2, \ldots, y_n)$$

The distance between x and y is

$$\sqrt{\sum_{i=1}^{n} (y_i - x_i)^2}$$

Sometimes, it is useful to think of the point x as a <u>vector</u>: this vector is just an arrow, drawn from the origin to the point. For the two-dimensional case, see figure 3.

Figure 3. A vector in two dimensions.

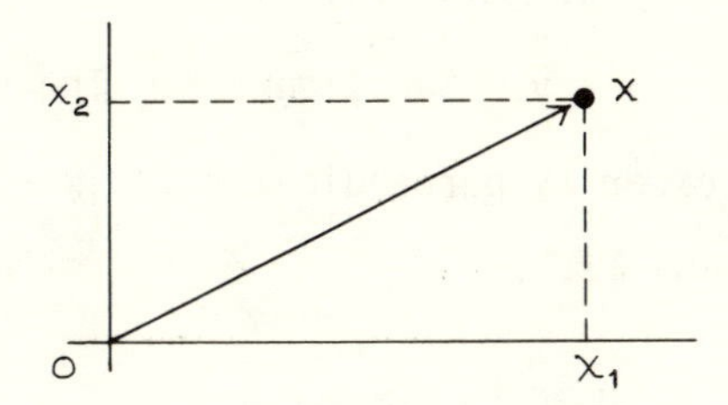

Now consider two vectors, x and y, in n-dimensional space. These two vectors lie on a plane, and can therefore be drawn as in figure 4. The <u>projection</u> of y on x is obtained by dropping a perpendicular from y onto x ; this gives the multiple of x closest to y. (Multiplying x by the number m, say, just means multiplying all the coordinates of x by m. This gives another point on the line joining the origin to x.)

Figure 4. Projecting y onto x.

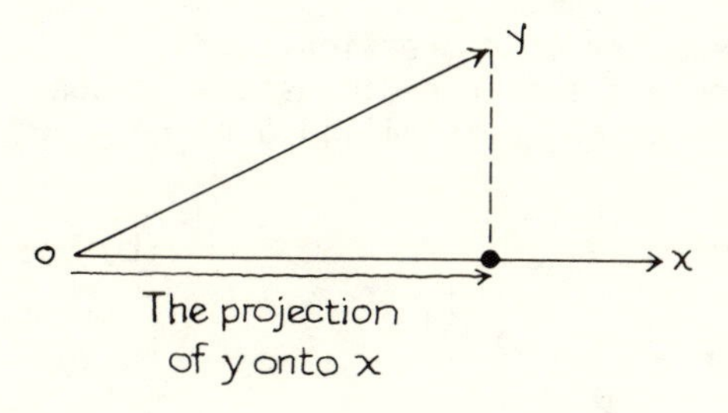

Regression involves two points in n-dimensional space. One point represents the deviations of x from its average $\bar{x}$:

$$x - \bar{x} = (x_1 - \bar{x}, x_2 - \bar{x}, \ldots, x_n - \bar{x})$$

Another is for the deviations of y from its average $\bar{y}$:

$$y - \bar{y} = (y_1 - \bar{y}, y_2 - \bar{y}, \ldots, y_n - \bar{y})$$

The regression of y on x can now be visualized as the projection of the vector $y - \bar{y}$ on the vector $x - \bar{x}$. Algebraically, $\hat{y} - \bar{y} = m(x - \bar{x})$ is the multiple of $x - \bar{x}$ closest to $y - \bar{y}$. See figure 5. The vector of residuals $e = (y - \bar{y}) - m(x - \bar{x})$ is necessarily perpendicular to $x - \bar{x}$. The algebraic counterpart is that, by 13.5 and 13.7,

$$\text{cov}(e,x) = \frac{1}{n} \sum_{i=1}^{n} e_i(x_i - \bar{x}) = 0$$

This is the technical backup to figure 1 on page 75. Notice that

- the SD of y is the length of the vector $y - \bar{y}$
- the SD of $\hat{y}$ is the length of the projection $\hat{y} - \bar{y}$
- the SD of e is the length of the residual vector e.

The squares of the lengths are the variances, and these add up by the Theorem of Pythagoras. Figure 5b below shows vectors; figure 1 on page 75 shows the lengths of these vectors.

Figure 5. The geometry of regression.
The vectors are in n-dimensional space,
but the triangle is all in a plane

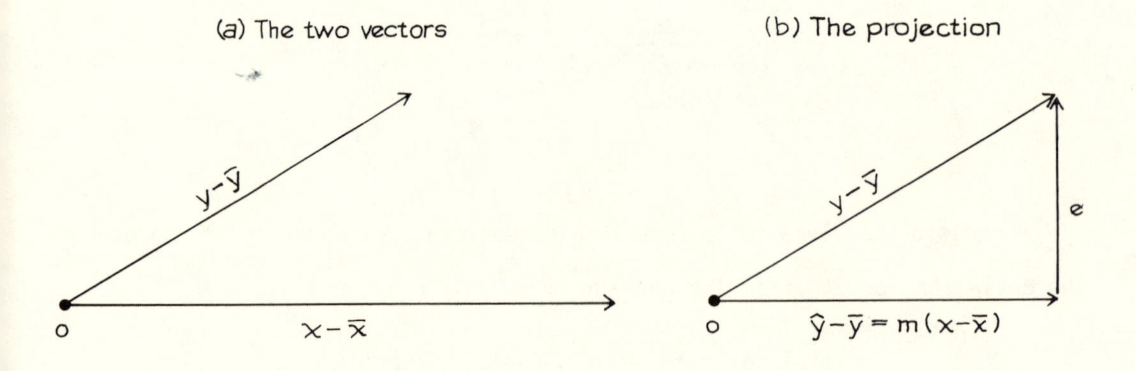

[There are no exercises to this section.]

PART B. THE ALGEBRA OF RANDOM VARIABLES

17. Random variables[1]

The object of this section is to introduce the somewhat formidable term random variable. The definition is as follows.

A random variable is a chance procedure for generating a number.

This definition will now be illustrated by a few examples. The first is somewhat artificial (see figure 1): Draw a ticket at random from the box, and take the number written on the ticket.

Figure 1. The random variable X .

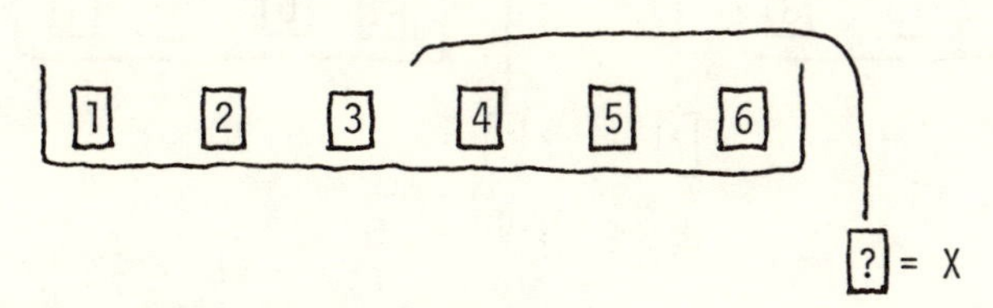

Here is another, more elaborate, example. Let X be as in figure 1. Let Y stand for the following procedure: Draw a ticket at random from another box, and take the number on that ticket.

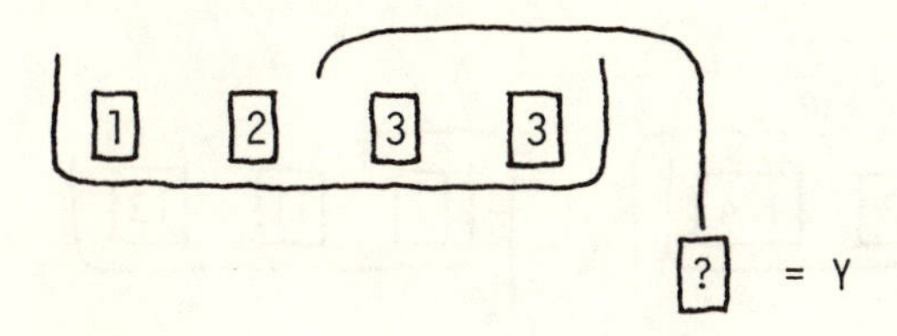

So Y is a second random variable. We can define a third random variable Z in terms of X and Y :

$$Z = X + Y$$

[1] Text, chapter 16.

In more detail, Z stands for this chance procedure: Draw a number at random from the first box; draw a number at random from the second box; add up the two numbers.

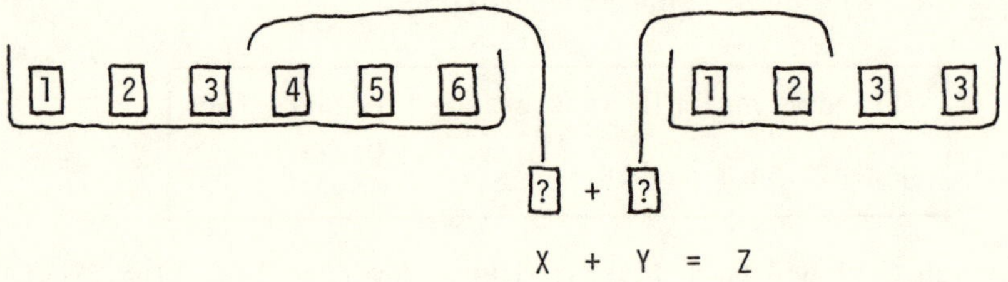

1 2 3 4 5 6 1 2 3 3

? + ?

$X + Y = Z$

Of course, many other random variables could be defined in similar ways. For instance,

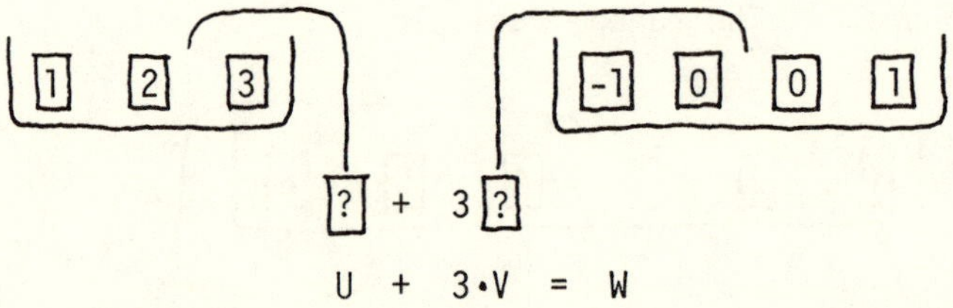

1 2 3 -1 0 0 1

? + 3 ?

$U + 3 \cdot V = W$

In more detail: Draw a number at random from the first box; draw a number at random from the second box, and multiply by 3; take the sum. Here the random variables are denoted by U, V, and W.

A final example:

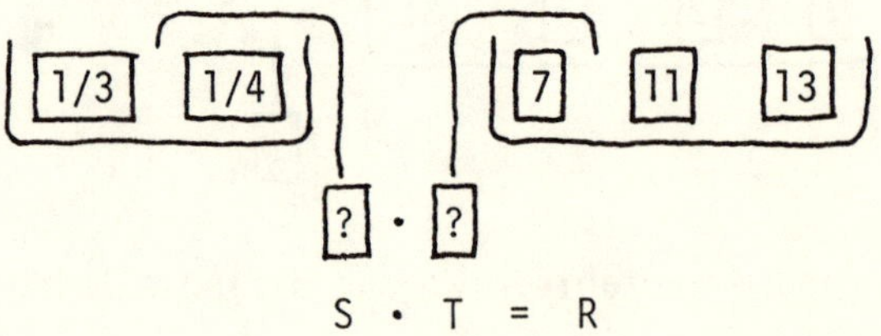

1/3 1/4 7 11 13

? · ?

$S \cdot T = R$

Here the draws are multiplied; the random variables are denoted by R, S, and T.

> An arithmetic operation on draws from boxes makes a random variable.

In the examples, the random variables were denoted by upper-case letters; X, Y, Z, etc. This is a convention of statistics.

There is a fussy but useful distinction between random variables and <u>observed values</u>. Let us go back to the first example

Here, X stands for the whole procedure: shake the box, draw out a ticket at random, take the number on this ticket. Suppose you went through all this, and got the number 4. This would be an <u>observed value</u>. A random variable is a procedure; an observed value is a result.

It is a convention of statistics to use lower case letters for observed values. Thus, x might be used to denote a typical observed value of the random variable X. Be careful, however: conventions are not always followed!

<u>Exercise set 17</u>

1. Someone draws a ticket at random from the box [1] [1] [2] [3] and gets [2]. What is the random variable? The observed value?

2. The following table shows observed values for three random variables. Fill in the blanks.

X	Y	2X + 3Y + 7
1	5	24
1	6	___
___	6	29
2	1	___

3. True or false, and explain: if X and Y are random variables, so is $X^2 + Y^2$.

4. Which observed value goes with which random variable? Are you sure of the match? Explain briefly.

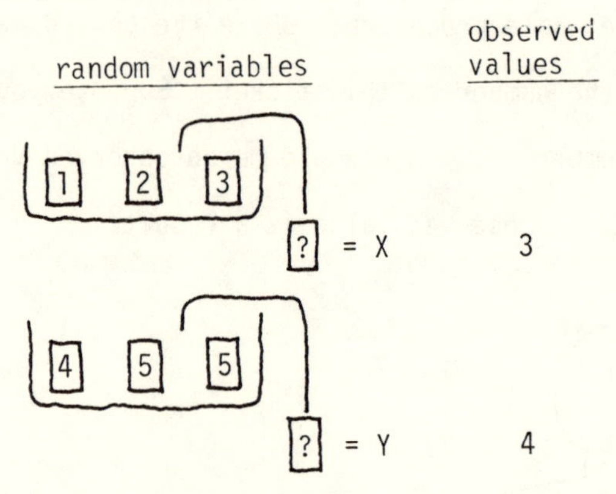

5. Repeat exercise 4, on the following.

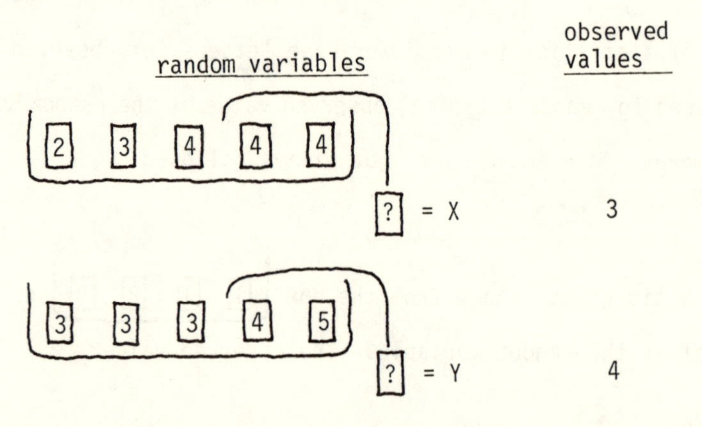

18. More random variables

Sometimes it is convenient to have a box where each ticket shows not one number but two. Drawing a ticket at random from such a box defines two random variables: the 1st number on the ticket, and the 2nd.

$$\text{1st number} = X$$

$$\text{2nd number} = Y$$

In this example, X stands for the following procedure: draw a ticket at random from the box, take the first number on the ticket.

What is the chance that X will be 1? This happens with two tickets out of 4, so the chance is two out of 4, or 50%. Statisticians write this, a trifle brutally, as

$$P(X = 1) = 50\%$$

The "P" means probability, or chance. And "X = 1" inside the parentheses means "X will be 1".

What is $P(2X + 3Y = 11)$? Here $2X + 3Y$ is a new random variable: draw a ticket at random from the box, take twice the first number plus three times the second number on the ticket.

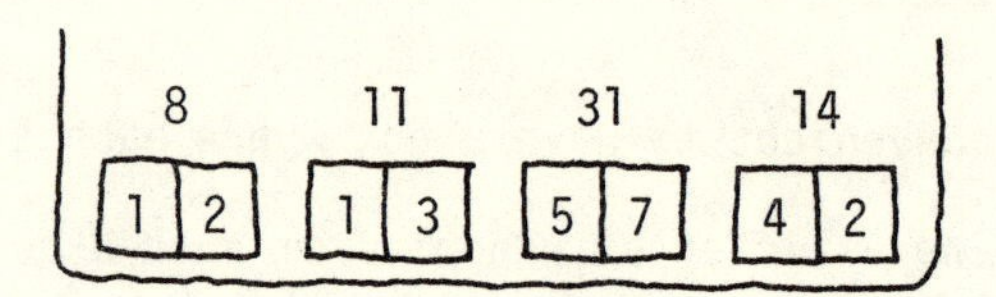

The figure shows the values of $2X + 3Y$ for each ticket. Only one ticket out of four has a value of 11, so

$$P(2X + 3Y = 11) = 25\%$$

Exercise set 18

1. Define U and V by the diagram below.

   (a) Explain, in words, what $U \cdot V$ means.

   (b) Find $P(U \cdot V = 3)$

   (c) Find $P(2U - 3V = 7)$

2. Repeat exercise 1, for the box below.

19. Independence[1]

Let X and Y be two random variables. They may be dependent or independent. To decide which, pretend that you know the value of X, and ask whether the chances for Y depend on that value. If so, X and Y are dependent. If the chances for Y do not depend on the value of X, then X and Y are independent. This is very brief, and some examples may help.

Example 1. Define X and Y by the diagram below. Are X and Y dependent or independent?

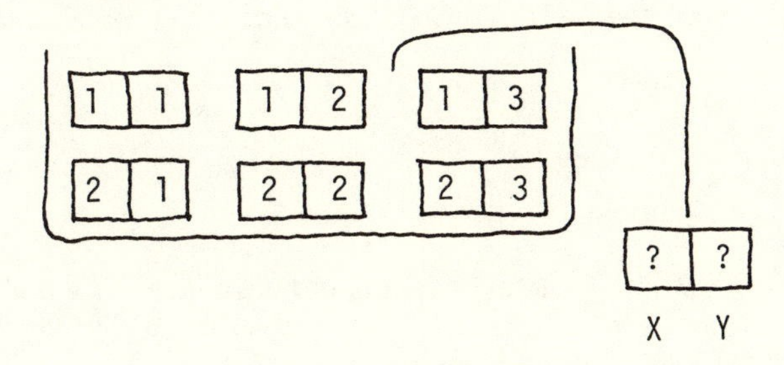

Solution. Pretend you know X is 1. Then Y is 1, 2 or 3 with one chance in three. Likewise if X is 2. The chances for Y do not depend on the value of X. So X and Y are independent.

Example 2. As in example 1, but with the following box:

Solution. This time, X and Y are dependent. For instance, if X is 1 then Y has two chances in four to be 3. But if X is 2, then Y has only one chance in four to be 3.

[1] Text, pp. 212-16

In this example, X and Y are dependent. But Y cannot be computed from X. (For instance, when X is 1, there are three ways Y could turn out; a formula isn't like that.) Sometimes the language of statistics is different from ordinary language.

Example 3. As in example 2, but somebody erases the first number on the ticket as it is drawn from the box.

Solution. The erasing business is irrelevant. The test for independence has nothing to do with whether you actually know the value of X. To make the test, you *pretend* to know the value. So, X and Y are dependent, just as in example 2.

Notes. (i) If X and Y are independent, so are Y and X: the relationship is symmetric.

(ii) In the diagrams on page 90 and elsewhere, with draws from two separate boxes, it is tacitly assumed that the draws are independent.

Exercise set 19

1. Define X and Y by the diagram below. Are X and Y dependent or independent? How about Y and X?

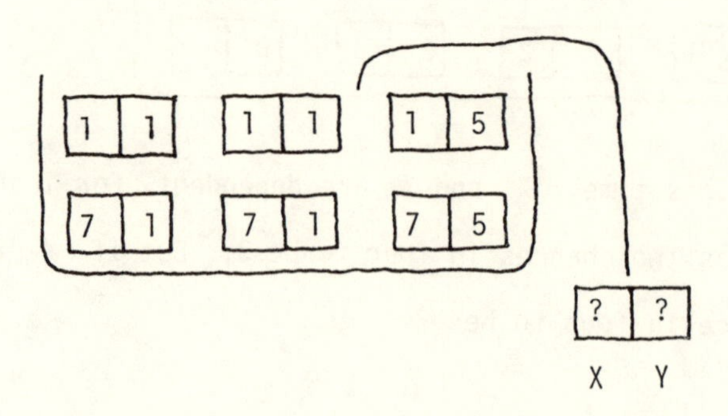

2. Repeat exercise 1, for the following box:

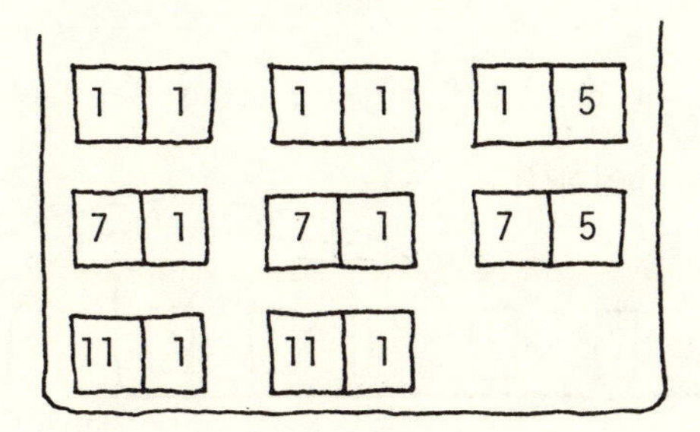

3. Repeat exercise 1, for the following box:

4. Repeat exercise 1, for the following box:

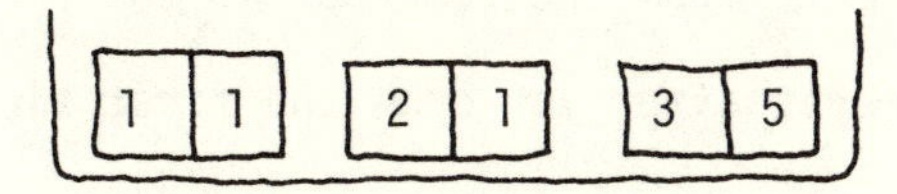

5. Repeat exercise 1, for the following box:

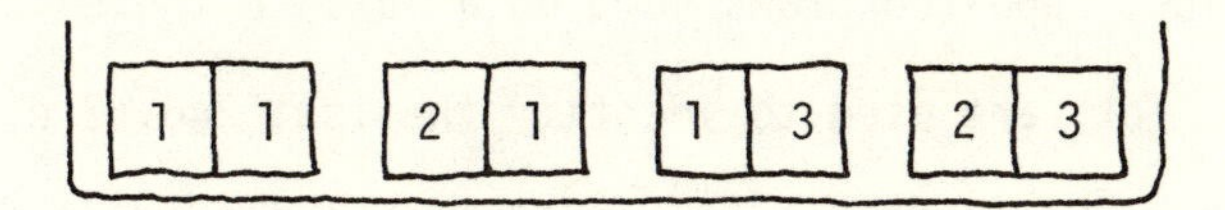

6. True or false, and explain: if you can't compute Y from X, or X from Y, then X and Y are independent.

7. True or false, and explain: if X and Y are independent, so are Y and X.

20. Repeated draws[1]

Here is a chance procedure for generating two numbers: make two draws at random from the box below.

The draws can be made either with or without replacement.

With replacement, you draw the first ticket at random from the box, make a note of the number on it, return the ticket to the box, and shake the box to mix up the tickets. Then you draw the second ticket at random. So the second draw is made from the same box as the first. On the other hand, without replacement, after you draw the first ticket, you set it aside and do not return it to the box. So the second draw is made from a box with one ticket less.

In statistics, repeated draws come up a lot, so there is some special notation: subscripts are used to identify the 1st, 2nd, 3rd,... draws. For instance, you can make 100 draws at random with replacement from the box below.

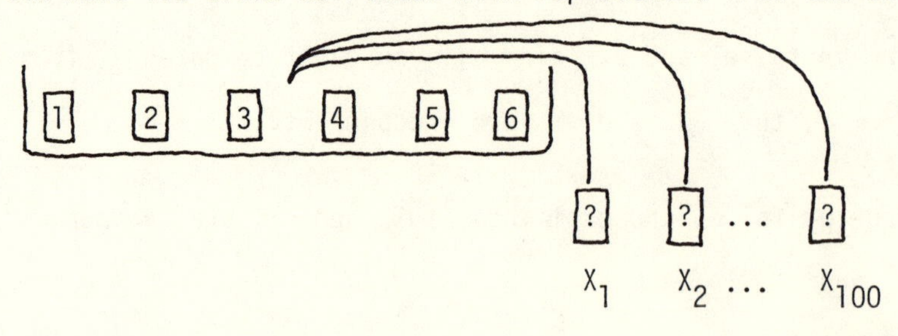

[1] Text, pp. 208ff and 212ff.

Then X_1 can be used to stand for the 1st draw, X_2 for the 2nd, X_{17} for the 17th, and X_{100} for the 100th. To be more explicit about the 17th, the random variable X_{17} stands for the following procedure: draw a hundred tickets at random with replacement from the box, take the number on the 17th draw.

The distinction between drawing with and without replacement ties into a previous distinction between independence and dependence. For instance, suppose you make two draws at random from a box:

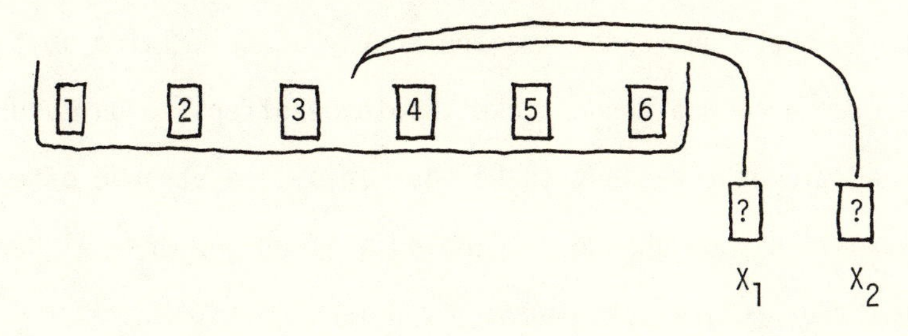

With replacement, the draws are independent. Even if you know how the first draw turned out, the chances for the second draw stay the same. Indeed, the second draw is made from exactly the same box as the first, because the first ticket gets replaced before the second draw is made.

By contrast, without replacement, the draws are dependent: Pretend you know how the first draw turned out; then the chances for the second draw depend on that result. For instance, if the first draw is 3, the chance for the second draw to be 6 is one in five. If the first draw is 6, however, the chance drops to zero.

The next example is not about independence, but may help in clarifying the language of statistics.

Example. Two draws are made at random without replacement from the box

Find the chance that the second draw will be 4.

Solution. The question itself may seem strange to many readers: To solve it, don't we need to know how the first draw turned out? To understand what the question means, imagine that you are walking up to the box. Before you even draw the first ticket, someone offers to pay you a dollar if the second ticket turns out to be 4. This is a strange offer, but worth something. What is your chance of winning, before you draw either ticket? That is what the problem asks for.

Fortunately, the solution is shorter than the explanation. There are twelve possible outcomes to the experiment, shown in the table below.

first draw \ second draw	1	2	3	4
1	no	1 2	1 3	1 4
2	2 1	no	2 3	2 4
3	3 1	3 2	no	3 4
4	4 1	4 2	4 3	no

Fon instance, the first draw could be 2 and the second draw 4; in the table, this is denoted by "2 4". Or, the first draw could be 3 and the second draw 4: in the table, this is "3 4" . However, if the first draw is 3, say, then the second draw cannot be 3: in the table, there is a "no" instead of a "3 3". All in all, out of the twelve possible outcomes in the table, three have a 4 on the second draw. So the chance is three out of twelve, or 1/4. This completes the solution.[1]

At this point, readers may be getting a bit nervous: What about three draws? Or suppose there are ten tickets in the box? It looks as if the details could easily get out of hand. Do not worry. We will be developing techniques which bypass such details. However, readers should do some calculations like the one in the example, in order to appreciate what happens later.

Exercise set 20

(This exercise set covers material from previous sections.)

1. A random variable is shown below:

[1] [2] [3] [3] → [?] = X

(a) $P(X = 1) = ?$

(b) Find the chance that X will be 1.

(c) $P(X = 3) = ?$

---

[1] Compare with p. 221 of the text. Readers may object to counting outcomes with their fingers. However, this method was good enough for Galileo: p. 222 of the text.

2. Two draws will be made at random with replacement from a box.

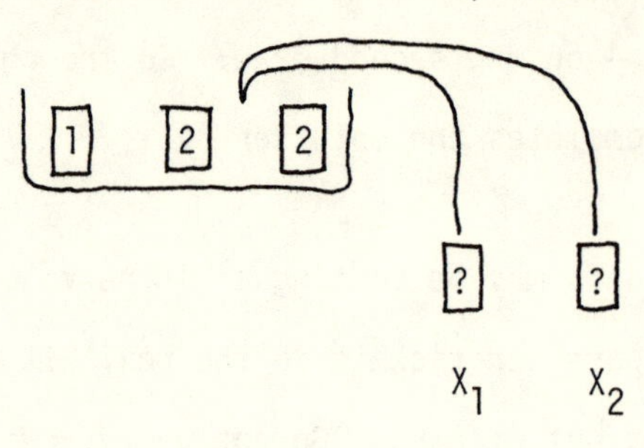

(a) $P(X_1 = 1) = ?$

(b) Find the chance that the second draw will be 2.

(c) Find the chance that the first draw will be 1, and the second draw will be 2.

(d) Is $P(X_1 = 1 \text{ and } X_2 = 2) = P(X_1 = 1) \cdot P(X_2 = 2)$?

(e) Are X_1 and X_2 independent?

3. Repeat exercise 2, if the draws are made without replacement.

4. Two draws are made at random with replacement from the box [1] [1] [2]. Let X_1 stand for the first draw, and X_2 for the second. True or false, and explain:

(a) "$X_1 + X_2$" and "$2 \cdot X_1$" are two different ways of writing the same random variable.

(b) $X_1 + X_2$ will be 2, 3, or 4.

5. A deck of cards is shuffled, and a five-card poker hand is dealt off the top of the deck. Is this like drawing at random with replacement? or without? Explain.

Technical notes. Let X_1 and X_2 be two draws made at random without replacement. In the example on page 100, it turned out that $P(X_1 = 4) = P(X_2 = 4)$. This is not a coincidence. Before either draw is made, the chances for the second draw must be the same as for the first draw. This is surprising, but it can be given a mathematical proof.

A relevant distinction can be drawn between conditional and unconditional probabilities. This is discussed in more advanced texts. In the same example, $P(X_2 = 4) = 1/4$ is an unconditional probability: it is computed with no condition on X_1. If $X_1 = 1$, say, then X_2 is 4 with one chance in three. This is a conditional probability: it is computed with a condition on X_1. The usual notation for a conditional probability is as follows:

$$P(X_2 = 4 | X_1 = 1) = 1/3$$

The relevant condition is stated after the vertical bar. The left-hand side of the display can be read as follows: the probability that $X_2 = 4$ given that $X_1 = 1$.

With respect to exercise 2d: if X and Y are independent random variables, i and j possible values, then

$$P(X = i \text{ and } Y = j) = P(X = i) \bullet P(Y = j)$$

21. A modelling issue

Part A of this booklet was about one kind of variable -- data variables. Part B is about another -- random variables. The two should be kept apart. A data variable is a list of numbers. A random variable is a chance procedure for generating a number.

Sometimes, a data variable can be viewed as a list of observed values of some random variables. The random variables are a model for the process which generated the data. This sort of modelling is basic to statistical inference: see parts VI and VII of the text, or part C of this booklet. Unfortunately, many numbers are not observed values of random variables, and this complicates the task of statistical inference.

Example. You go out tomorrow morning and ask the age of the first person you meet. Is this a random variable?

Discussion. We think not (although the issue is somewhat controversial). There is no clear way to assign chances to the various possibilities.

[There are no exercises to this section.]

22. Distribution tables

The distribution table of a random variable shows the possible values, and the chances of getting them. For instance, take the following random variable:

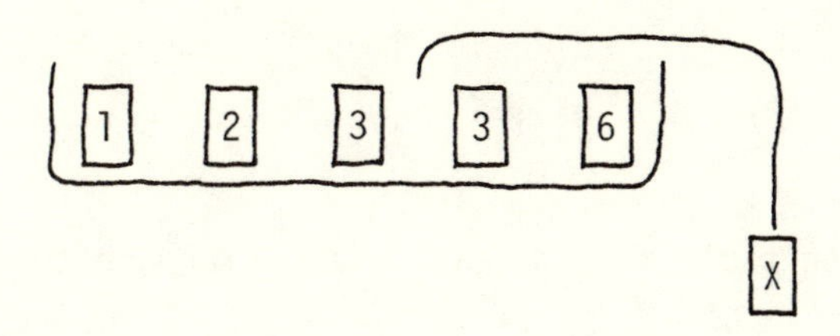

There are four possible values: 1, 2, 3, 6. The distribution table is as follows:

value	chance
1	1/5
2	1/5
3	2/5
6	1/5

Example. Find the distribution table of X. Find the distribution table of X + Y. The random variables X and Y are defined in the figure below.

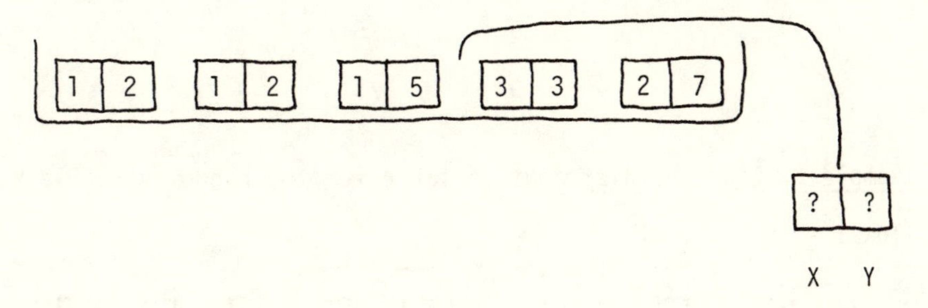

Solution. The possible values for X are 1, 2, and 3. Three tickets out of five have a "1" in the first position, so there are three chances in five that X will be 1. The other values can be dealt with in a similar way.

Distribution table for X

value	chance
1	3/5
2	1/5
3	1/5

To get the distribution table for $X + Y$, just go through the box and write $X + Y$ on each ticket:

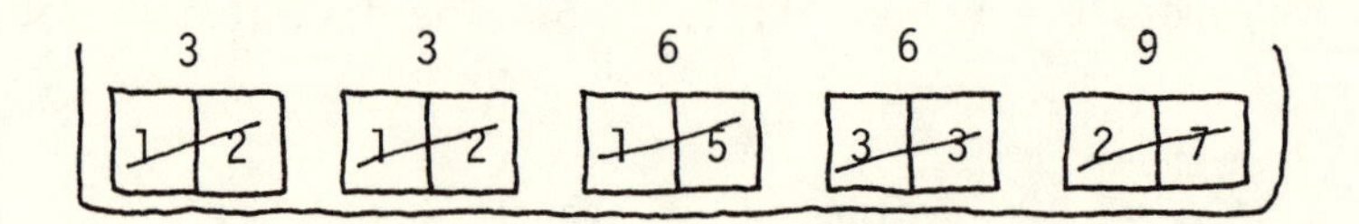

The possible values for the sum are 3, 6, and 9. Two tickets out of five show a sum of 3, so there are two chances out of five that the sum will be 3. The other values can be dealt with in a similar way.

Distribution table for X + Y

value	chance
3	2/5
6	2/5
9	1/5

Example 2. Find the distribution table for the random variable W, defined as follows:

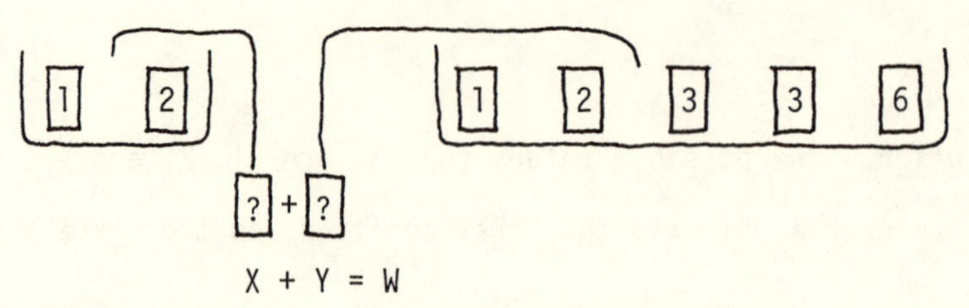

Solution. The key is the diagram below.

		Y: 1	2	3	3	6
X	1	11	12	13	13	16
	2	21	22	23	23	26

In short, X and Y are like the 1st and 2nd number on one ticket drawn at random from the following box:

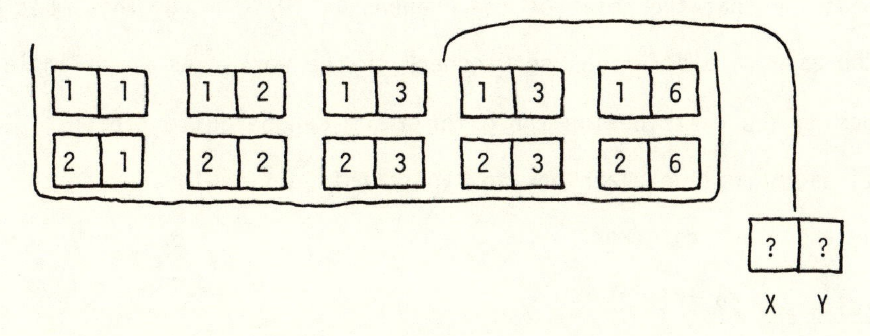

Now the method of example 1 can be used: just go through the box, and write X + Y on each ticket;

2	3	4	4	7
3	4	5	5	8

The possible values for X + Y are: 2, 3, 4, 5, 7, 8. Take the value 4, say. Since three tickets out of ten show a sum of 4, there are three chances in ten for X + Y to be 4. Each of the possible values can be dealt with in the same way.

Distribution table for X + Y

value	chance
2	1/10
3	2/10
4	3/10
5	2/10
7	1/10
8	1/10

Notice that the total of the chances is 10/10 = 1. This must always be the case, and offers a useful check on the work. As the example indicates, computing the distribution table for a sum can be quite tedious. In later sections, we will explain how to bypass this difficulty, using the expected value and standard error.

Exercise set 22

(This exercise set covers material from previous sections.)

1. Find the distribution table for the random variable Z shown below.

1 2 2 0 1

3• X + 4• Y = Z

2. Find the distribution table for the random variable W shown below.

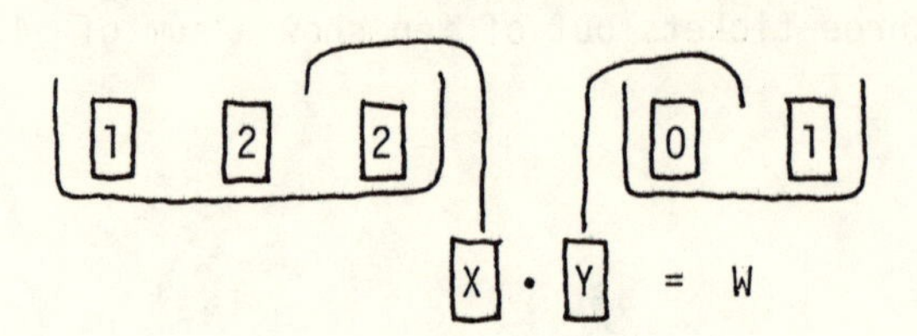

Hint: this is like example 2 on page 106, with multiplication instead of addition.

3. Find the distribution table for the random variable S shown below: the draws are made without replacement.

Hint: see the example on page 100.

4. Random variables X and Y are shown below.

   a) Are X and Y independent?

   b) Find the distribution table for 2X - Y.

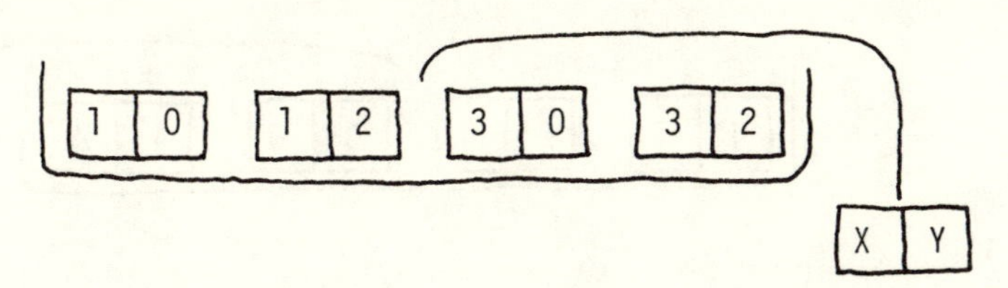

5. Repeat exercise 4, the random variables being as follows:

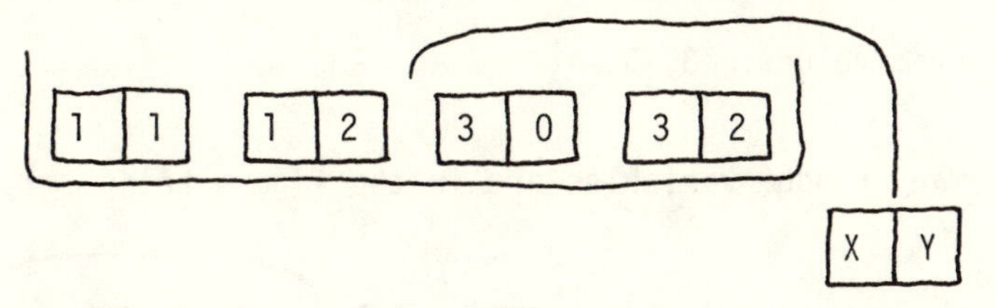

6. You will be paid whatever some random variable turns out to be, in dollars. Which random variable do you prefer?

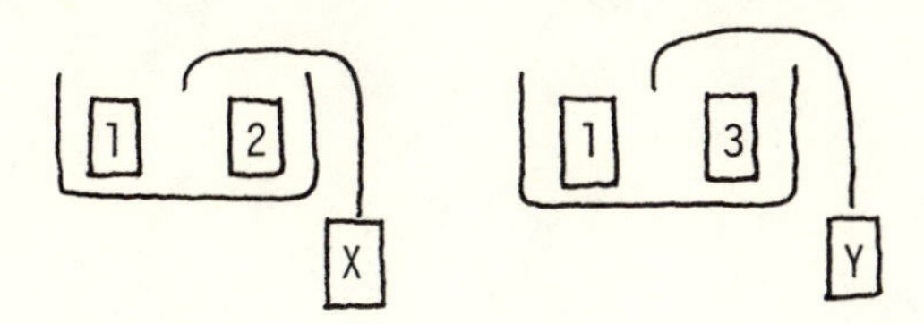

7. Repeat exercise 6, choosing between U and V:

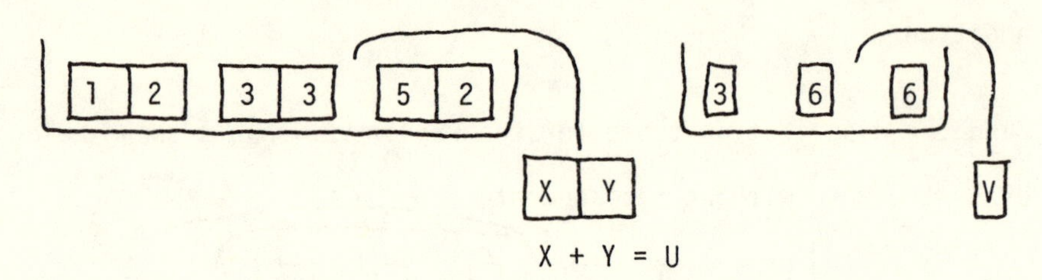

8. Find P(X + Y = 7):

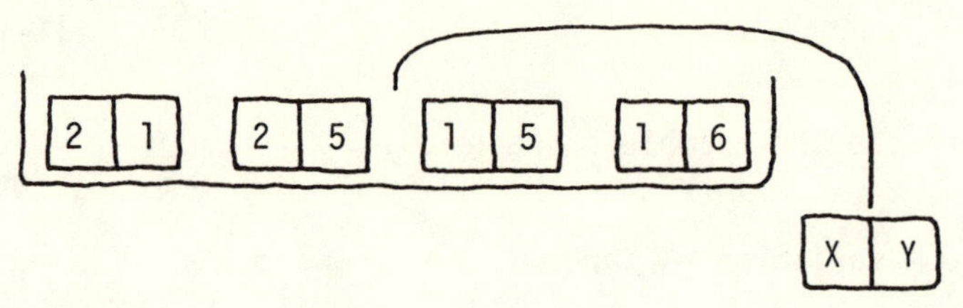

9. Match the random variables U and V with the observed values. Are you sure of the match?

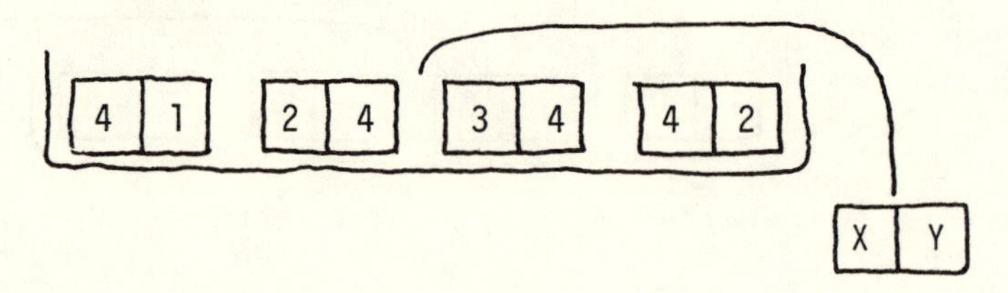

Random variables: U = X + Y, V = X - Y

Observed values: 3, 5.

10. How many random variables are in the figure below: 1, 3, or 5?

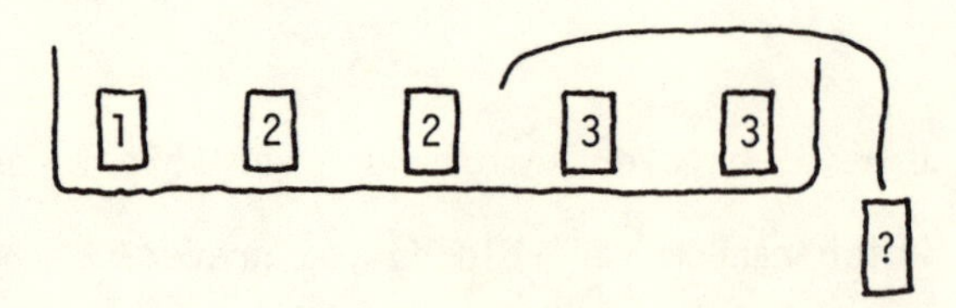

23. Expected value[1]

Let X be a random variable. Its expected value is denoted as follows: E(X). This is just shorthand: The "E" stands for "expected value", and the parentheses for "of". So E(X) can be read "E of X": note that the E does not multiply the X.

The expected value of a random variable is a number. And, as the name suggests, the random variable will be somewhere around this number:

$$X = E(X) + \text{chance error}$$

This section and the next will explain how to compute E(X). The chance error will be dealt with later.

> If X is a draw from a box of numbered tickets, then E(X) can be computed by adding up the values and dividing by how many tickets there are in the box.

Example 1. Define X by the diagram below. Find E(X).

[1] [2] [3] [3] [6]

[?] = X

Solution.

$$E(X) = \frac{1 + 2 + 3 + 3 + 6}{5} = 3$$

---

[1]Text, pp. 255-256. "Expected value" is also called "expectation" or "mathematical expectation".

<u>Example 2</u>. Define X as in example 1. Let $Y = X^2$, the square of X. Find E(Y).

<u>Solution</u>. The first step is to go through the box and write the value of Y on each ticket:

1	4	9	9	36
~~1~~	~~2~~	~~3~~	~~3~~	~~6~~

Now use the old method:

$$E(Y) = \frac{1 + 4 + 9 + 9 + 36}{5} = \frac{59}{5} = 11.8$$

<u>Example 3</u>. Suppose X and Y are as follows:

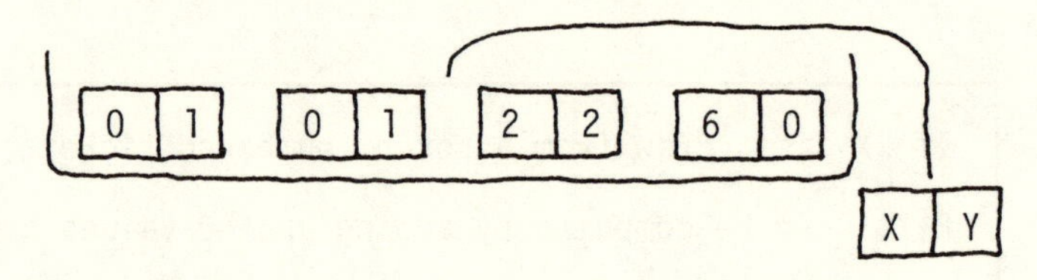

(a) E(X) = ?

(b) E(Y) = ?

(c) E(2X + 3Y) = ?

(d) Does E(2X + 3Y) = 2E(X) + 3E(Y)?

<u>Solution</u>. For a), just add up the values of X on the tickets, and divide by 4, the number of tickets:

$$E(X) = \frac{0 + 0 + 2 + 6}{4} = 2$$

Likewise for part b), using the values of Y:

$$E(Y) = \frac{1 + 1 + 2 + 0}{4} = 1$$

For part c), write the value of $2X + 3Y$ on each ticket, and then use the same method:

3	3	10	12
0 1	0 1	2 2	6 0

$$E(2X + 3Y) = \frac{3 + 3 + 10 + 12}{4} = \frac{28}{4} = 7$$

For part d), the equation is true: $7 = 2 \cdot 2 + 3 \cdot 1$

The expected value of a random variable can also be computed from the distribution table, as follows:

> The expected value is the sum of the products "value x chance".

For instance, suppose X has this distribution table:

value	chance
2	2/5
3	2/5
5	1/5

Then

$$E(X) = 2 \cdot \frac{2}{5} + 3 \cdot \frac{2}{5} + 5 \cdot \frac{1}{5} = \frac{15}{5} = 3$$

Example 4. Define X and Y as in example 3 above. Let $U = 2X + 3Y$. Find $E(U)$ from the distribution table.

Solution. The box for U is as follows:

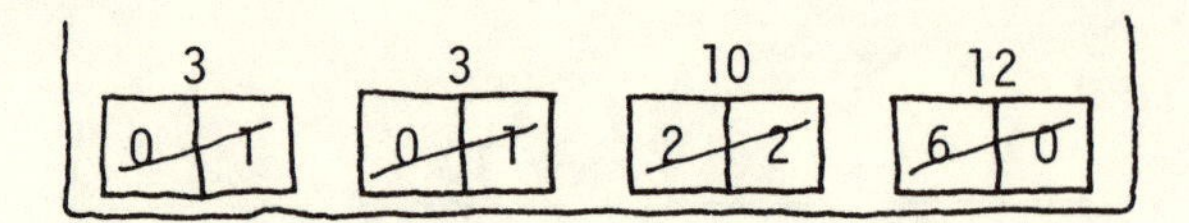

The possible values for U are 3, 10, and 12.

Distribution table for U

value	chance
3	2/4
10	1/4
12	1/4

So

$$E(U) = 3 \cdot \frac{2}{4} + 10 \cdot \frac{1}{4} + 12 \cdot \frac{1}{4} = \frac{28}{4} = 7$$

This agrees with the answer found before, in part c) of example 3.

Exercise set 23

1. True or false, and explain:

   (a) $E(X) = E \cdot X$

   (b) $E(X) = X(E)$

   (c) $E(X)$ is the observed value of X

   (d) $E(X)$ is a possible value of X

   (e) $E(X)$ is the most likely value of X

   (f) $E(X)$ is a number

   (g) $P(X = 3) = P \cdot (X = 3)$

   (h) $P(X = 3)$ simplifies to $P(X) = P(3)$

   (i) $P(X = 3)$ is a number between 0 and 1

2. Make up a random variable with the distribution table given below; find its expected value by the two methods of this section.

value	chance
-2	1/5
0	1/5
1	1/5
3	2/5

3. The possible values for a certain random variable are 1, 2, 3, and 7. Part of its distribution table is given below. Fill in the blank, and find the expected value.

value	chance
1	3/10
2	1/10
3	____
7	4/10

4. Random variables X and Y are defined below. Find E(X), E(Y), and E(4X - Y + 3) by the two methods of this section. Is E(4X - Y + 3) = 4E(X) - E(Y) + 3?

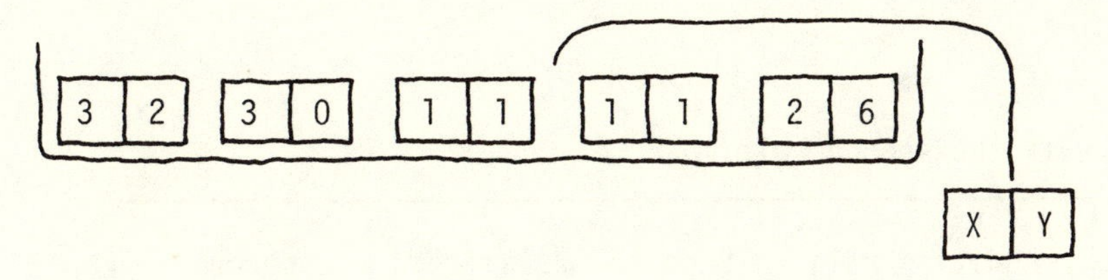

24. More on expected value

The methods of the last section are basic, but sometimes hard to use. The object in this section is to present three rules which will enable you to compute many expected values very efficiently. Here is the first rule.

> If X is a single draw from a box of numbered tickets, then
>
> $$E(X) = \text{average of box}$$

"Average of the box" is an abbreviation for the average of the numbers in the box, so we just have the method on page 111, in different language.

Suppose, for instance, that X is defined as follows:

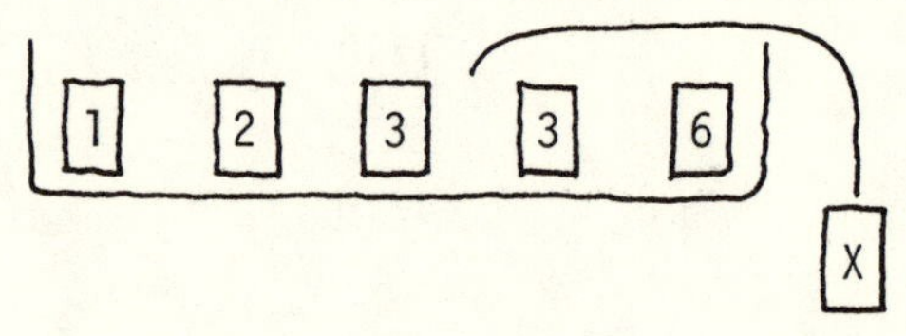

Then

$$\text{average of box} = \frac{1 + 2 + 3 + 3 + 6}{5} = 3$$

So

$$E(X) = 3$$

Here is the second rule.

> Addition rule. Suppose X and Y are random variables, while a, b, and c are constants. Then
>
> $$E(aX + bY + c) = aE(X) + bE(Y) + c$$

This rule applies as well to three or more random variables.

Example 1. Find E(W), where W is defined as follows:

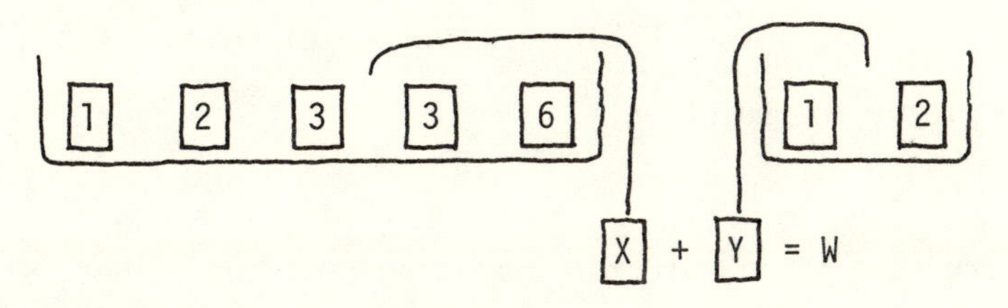

Solution. First compute E(X) and E(Y):

$$E(X) = \frac{1 + 2 + 3 + 3 + 6}{5} = \frac{15}{5} = 3$$

$$E(Y) = \frac{1 + 2}{2} = \frac{3}{2} = 1.5$$

Now use the addition rule:

$$E(W) = E(X + Y) = E(X) + E(Y) = 3 + 1.5 = 4.5$$

This completes the solution.

Example 2. Suppose n draws are made at random with replacement from a box of numbered tickets. Let S be the sum of the draws. Show that[1]

$$E(S) = n\cdot\text{average of box}$$

Solution. Let X_1 be the 1st draw, X_2 the 2nd draw, and so on up to X_n, the nth draw. Then

$$S = X_1 + X_2 + \dots + X_n$$

By the addition rule,

$$E(S) = E(X_1) + E(X_2) + \dots + E(X_n)$$

There are n terms in this sum; and each equals the average of the box.

[1]Text, p. 256

So

$$E(S) = n \cdot \text{average of box}$$

This completes the solution.

There is a third rule for computing expected values, which is sometimes useful:

<u>Multiplication rule</u>. If X and Y are independent, then $$E(X \cdot Y) = E(X) \cdot E(Y)$$

If X and Y are dependent, then $E(X \cdot Y)$ may differ from $E(X) \cdot E(Y)$; the multiplication rule does not apply.

<u>Example 3</u>. A die is thrown twice. Let X_1 be the number of spots on the 1st throw; X_2 the number on the 2nd throw. Find $E(X_1 \cdot X_2)$.

<u>Solution</u>. First,

$$E(X_1) = E(X_2) = \frac{1+2+3+4+5+6}{6} = \frac{7}{2}$$

Since X_1 and X_2 are independent, the multiplication rule can be used:

$$E(X_1 \cdot X_2) = E(X_1) \cdot E(X_2) = \frac{7}{2} \cdot \frac{7}{2} = 12\tfrac{1}{4}$$

<u>Exercise set 24</u>

1. Define random variables U and V as follows:

$$U = X + Y, \qquad V = X \cdot Y$$

(a) Find E(U).

(b) Does the addition rule apply?

(c) Find E(V).

(d) Does the multiplication rule apply?

2. Define W by the diagram below. Find E(W).

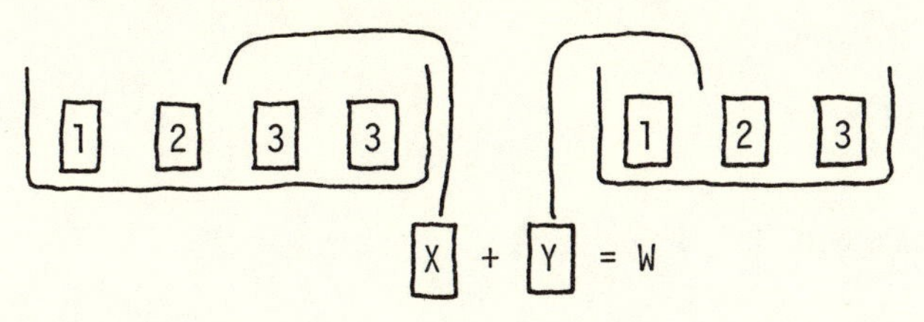

3. Let W be as follows:

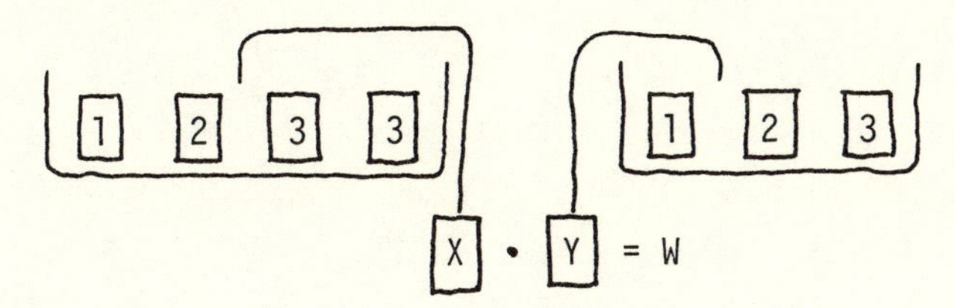

(a) Find E(W).

(b) The random variable W is a chance procedure for getting a number. Explain carefully what this procedure is.

4. Let X_1 and X_2 be two draws made at random without replacement from the box

1 2 3 4

(a) $E(X_1) = ?$

(b) $E(X_2) = ?$ Or do you need to know X_1?

(c) $E(X_1 + X_2) = ?$

(d) Does the addition rule apply?

(e) $E(X_1 \cdot X_2) = ?$

(f) Does the multiplication rule apply?

5. Let X and Y be random variables; x and y, data variables; a, b, and c, constants. True or false, and explain:

   (a) $E(aX + bY) = aE(X) + bE(Y)$

   (b) $E(ax + bY) = aE(x) + bE(Y)$

   (c) $E(xY) = xE(Y)$

   (d) $E(Y+c) = E(Y) + c$

   (e) $E(X \cdot Y) = E(X) \cdot E(Y)$

6. This continues example 2 on page 117. Let $\bar{X} = S/n$, so $\bar{X}$ is the average of the draws.

   (a) Is $\bar{X}$ a random variable?

   (b) If so, what is $E(\bar{X})$?

25. Variance

In section 23, we saw that

(1) $X = E(X) + \text{chance error}$

The likely order of magnitude of the chance error is given by a number called the <u>standard error</u>, abbreviated SE.

> The random variable X will be around $E(X)$, give or take an SE or so.[1]

For instance, if $E(X) = 100$ and the SE is 10, then X is likely to be in the range from $100 - 10 = 90$ to $100 + 10 = 110$. A value of 153, say, is very unlikely: over 5 SEs away from the expected value.

The SE is defined in terms of the variance:

(2) $\text{SE of } X = \sqrt{\text{var } X}$

Variance is an important technical concept, but without much intuitive meaning. To interpret variance, take the square root: that gives the SE.

By definition,

(3) $\text{var } X = E\{[X - E(X)]^2\}$

This is a complicated formula, let us take it slowly:

- X is a random variable;
- $E(X)$ is the expected value of X;
- $X - E(X)$ is the deviation of X from its expected value: this is the chance error in equation (1);

[1] Text, pp. 257ff. We use "SE" for random variables and "SD" for lists, but this distinction is not conventional.

- $[X - E(X)]^2$ is the square of the chance error;
- var X is the expected value of the square of the chance error.

For an example on equation (3), suppose X is as follows:

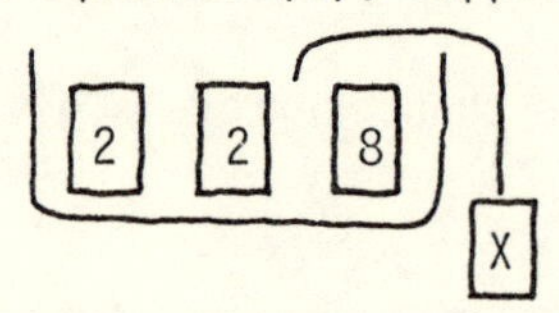

Then

$$E(X) = \frac{2 + 2 + 8}{3} = 4$$

To use equation (3), go through the box and write $[X - E(X)]^2$ on each ticket:

$(2-4)^2$ $(2-4)^2$ $(8-4)^2$

2 2 8

Now

$$\text{var } X = E\{[X - E(X)]^2\}$$

$$= \frac{(2-4)^2 + (2-4)^2 + (8-4)^2}{3}$$

$$= \frac{24}{3}$$

$$= 8$$

This completes the calculation of var X by formula (3). The interpretation: a draw from the box will be around 4, the expected value; it will be off 4 by $\sqrt{8} \approx 3$ or so, $\sqrt{8}$ being the standard error. This random variable is so simple that computing the expected value and standard error may seem more trouble than it's worth. The usefulness of these two concepts will be clear later.

Equation (3) may be hard to use, so we give some shortcut rules for computing variances. The first just restates equation (3) in simpler language, using "var of box" as an abbreviation for the variance of the numbers in the box.

> If X is a single draw from a box of numbered tickets, then
>
> $$\text{var } X = \text{var of box}$$

Example 1. Define X as follows:

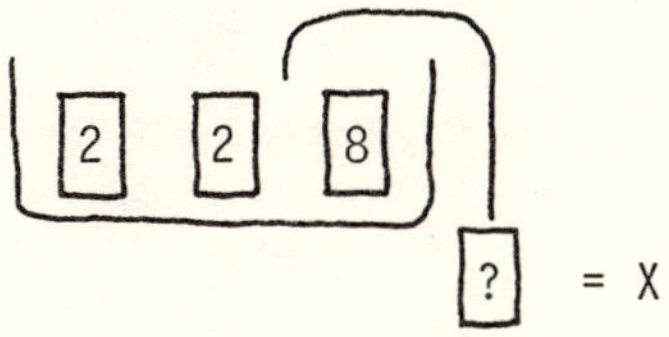

2 2 8

? = X

Find var X.

Solution. First,

$$\text{ave of box} = \frac{2+2+8}{3} = 4$$

$$\text{var of box} = \frac{(2-4)^2 + (2-4)^2 + (8-4)^2}{3} = \frac{24}{3} = 8$$

By the rule, var X is 8, completing the solution. Compared with the work on the previous page, the arithmetic is the same: only the language is different.

> Addition rule for variances. If X and Y are independent random variables, while a, b, and c are constants, then
>
> $$\text{var}(aX + bY + c) = a^2 \cdot \text{var } X + b^2 \cdot \text{var } Y$$

This rule does not apply to dependent variables. However, it can be extended to three or more independent variables.

The addition rule is complicated, so let us take some special cases. First, take $a = 1$ and $b = 0$. The rule says

(4) $\quad \text{var}(X + c) = \text{var}\, X$

In other words, adding a constant c to every value of the random variable X does not change the variance. This is because the spread around the expected value does not change.

Next, take $b = 0$ and $c = 0$. The rule says:

(5) $\quad \text{var}(aX) = a^2 \cdot \text{var}\, X$

This will make more sense in terms of standard errors: suppose $a = 10$, and take square roots in (5):

$$\text{SE of } 10X = 10 \cdot \text{SE of } X$$

In other words, multiplying every value of the random variable X by the constant 10 multiplies the SE by 10. That is because the spread around the expected value gets multiplied by 10.

Finally, take $a = 1$ and $b = 1$ and $c = 0$:

(6) If X and Y are independent random variables, then

$$\text{var}(X + Y) = \text{var}\, X + \text{var}\, Y$$

In other words, the variance of a sum of independent random variables is the sum of the variances. This is the heart of the addition rule.

Example 2. Find $E(T)$ and var T, where

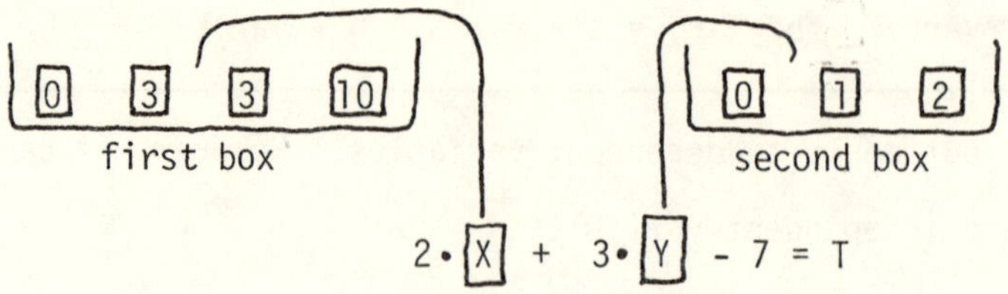

first box: 0 3 3 10

second box: 0 1 2

$$2 \cdot \boxed{X} + 3 \cdot \boxed{Y} - 7 = T$$

Solution. We do the arithmetic for the first box:

$$\text{ave of first box} = \frac{0 + 3 + 3 + 10}{4} = 4$$

$$\text{var of first box} = \frac{(0-4)^2 + (3-4)^2 + (3-4)^2 + (10-4)^2}{4} = \frac{54}{4} = 13\tfrac{1}{2}$$

The arithmetic for the second box is similar:

$$\text{ave of second box} = 1$$

$$\text{var of second box} = \frac{2}{3}$$

Now $E(T)$ can be computed:

$$\begin{aligned} E(T) &= 2E(X) + 3E(Y) - 7 \\ &= 2 \cdot 4 + 3 \cdot 1 - 7 \\ &= 4 \end{aligned}$$

Also, var T can be computed by the addition rule:

$$\begin{aligned} \text{var } T &= 2^2 \cdot \text{var } X + 3^2 \cdot \text{var } Y \\ &= 4 \cdot 13\tfrac{1}{2} + 9 \cdot \tfrac{2}{3} \\ &= 60 \end{aligned}$$

This completes the solution. The interpretation: T will be around 4, give or take $\sqrt{60} \approx 8$ or so. The give-or-take number may seem large, but remember that T could be as small as -7, or as large as 19, and these extremes are not unlikely.

Example 3. Let T be the sum of n draws made at random with replacement from a box of numbered tickets. Show that the standard error of T is given by the formula

$$\sqrt{n} \cdot \text{SD of box}$$

Solution. Let X_1 be the 1st draw, X_2 the 2nd, and so on up to X_n, the nth draw. Then the sum of the draws is

$$T = X_1 + X_2 + \ldots + X_n$$

Since the draws are made with replacement, they are independent. So

$$\text{var } T = \text{var } X_1 + \text{var } X_2 + \ldots + \text{var } X_n$$

There are n terms in this sum, and each is equal to var of box. Thus,

$$\text{var } T = n \bullet \text{var of box}$$

Taking square roots,

$$\begin{aligned} \text{SE of } T &= \sqrt{\text{var } T} \\ &= \sqrt{n} \bullet \sqrt{\text{var of box}} \\ &= \sqrt{n} \bullet \text{SD of box} \end{aligned}$$

This completes the solution. The formula in example 3 is called the square root law.[1]

Example 4. A hundred draws are made at random with replacement from the box

1	2	3	4	5	6

The sum of the draws will be around ______, give or take ______ or so.

Solution. First,

$$\text{ave of box} = \frac{1 + 2 + 3 + 4 + 5 + 6}{6} = 3\tfrac{1}{2}$$

Using formula 7.3 on page 34,

[1] Text, p. 258.

$$\text{var of box} = \frac{1^2 + 2^2 + 3^2 + 4^2 + 5^2 + 6^2}{6} - (3\tfrac{1}{2})^2 = \frac{35}{12} \approx 3$$

The expected value of the sum is

$$n \cdot \text{ave of box} = 100 \cdot 3\tfrac{1}{2} = 350$$

The variance of the sum is

$$n \cdot \text{var of box} \approx 100 \cdot 3 = 300$$

The SE of the sum of $\sqrt{300} \approx 17$. So the sum will be around 350, give or take 17 or so.[1]

Terminology.

The same term "variance" is used in two different (but similar) senses:

- As applied to a list of numbers. If x is a data variable, then var x equals the average square deviation from the average, namely, $\frac{1}{n} \sum_{i=1}^{n} (x_i - \bar{x})^2$.
- As applied to a random variable. If X is a random variable, then var X equals the expected square deviation from the expected value, namely, $E\{[X - E(X)]^2\}$.

In each case, you have to think whether "var" is used in the data sense, or the random variable sense. Usually, this will be clear from the context.

Exercise set 25

(This exercise set covers material from previous sections.)

1. If $E(X) = 1000$ and var $X = 100$, then X will be around ______ give or take _____ or so.

---

[1] See chapter 17 of the text.

2. A hundred draws will be made at random with replacement from the box [1] [2] [3] [3] [6]. The sum of the draws will be around ________, give or take ________ or so.

3. True or false, and explain:

   (a) Adding 3 to every value of a random variable adds 3 to the expected value.

   (b) Adding 3 to every value of a random variable adds 3 to the standard error.

   (c) Doubling every value of a random variable doubles the expected value.

   (d) Doubling every value of a random variable doubles the variance.

   (e) If X and Y are independent, SE of $(X + Y)$ = SE of X + SE of Y

   (f) If X and Y are independent, var $(X + Y)$ = var X + var Y

4. Ten draws are made at random with replacement from a box of numbered tickets. Let $X_1, X_2, \ldots, X_{10}$ denote these draws.

   (a) Explain carefully what X_7 means. What does the 7 refer to?

   (b) Are $X_1+X_2+\ldots+X_{10}$ and $10X_1$ different formulas for the same variable?

   (c) SE of $(X_1+X_2+\ldots+X_{10}) = 100 \cdot$ SE of X_1? Or $10 \cdot$ SE of X_1? Or $\sqrt{10} \cdot$ SE of X_1? Explain.

   (d) var $(X_1+X_2+\ldots+X_{10}) = 100 \cdot$ var X_1? Or $10 \cdot$ var X_1? Or $\sqrt{10} \cdot$ var X_1? Explain.

5. Define X and Y as shown below.

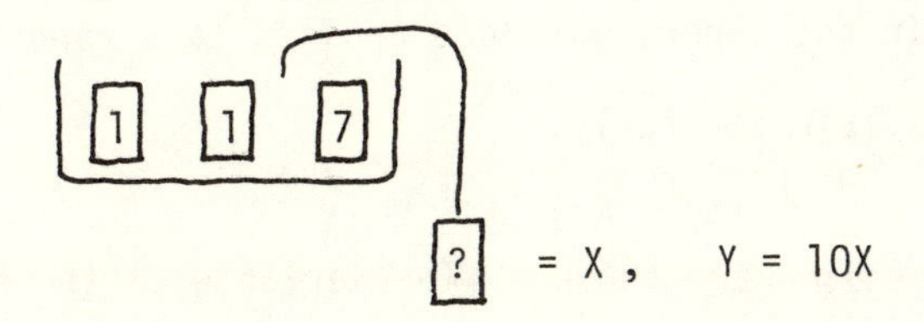

(a) The random variable Y is a chance procedure for generating a number. Explain carefully what this procedure is.

(b) SE of Y = 100 • SE of X? Or 10 • SE of X? Or $\sqrt{10}$ • SE of X?

(c) var Y = 100 • var X? Or 10 • var X? Or $\sqrt{10}$ • var X?

(d) P(Y = 10) = ?

6. Five draws will be made at random without replacement from the box [1] [2] [2] [3] [3] . The sum of the draws will be around ________, give or take ________ or so. Does the square root law apply? Explain your reasoning.

7. Random variables X and Y are given below. Find var X, var Y, and var (X + Y). Does equation (6) apply?

Hint: let U = X + Y; to find var U, write the values of U on each ticket, and proceed as on page 122.

8. The word "variance" is used in two sentences below. In one, variance applies to data. In the other, variance applies to a random variable. Which is which? Explain carefully.

   (i) A pair of dice are thrown. The variance of the number of spots is about 6.

   (ii) Among the 6,672 persons in cycle I of the Health Examination Survey, the variance of the number of years of schooling completed is about 10.

9. Example 1 on page 123 is about a random variable X. What is the difference between X and the list of numbers in the box [2] [2] [8] ?

10. Example 2 on page 124 is about a random variable T. Explain carefully the chance procedure that T stands for.

Technical notes

In the informal discussion, we do not distinguish between "standard error" and "probable error"; and we ignore problems of interpretation that may be created by bimodality, skewness, or long tails.

Suppose X and Y are independent random variables, c is a constant. Since var (X + c) = var X, readers may conclude that var (X + Y) should equal var X. This is a mistake, although a natural one. The point is that adding the constant c to X does not change the variability. However, adding the random variable Y to X does increase the variability.

26. More on variance

In this section, we will explain why the addition rule for variances (page 123) is valid. We will also explain what happens if X and Y are dependent. However, a new concept is needed: covariance. This concept has no direct interpretation. Basically, covariances help in computing variances, and variances help in computing standard errors. It is only the standard error which has a direct interpretation (page 121).

Let X and Y be random variables. By definition,

$$(1) \qquad \text{cov}(X,Y) = E\{[X - E(X)] \cdot [Y - E(Y)]\}$$

In words, $\text{cov}(X,Y)$ is the expected value of a product: the first factor is the chance error in X; the second, the chance error in Y.

An alternative formula:

$$(2) \qquad \text{cov}(X,Y) = E(X \cdot Y) - E(X) \cdot E(Y)$$

To derive (2) from (1), expand the product:

$$[X - E(X)] \cdot [Y - E(Y)] = X \cdot Y - X \cdot E(Y) - E(X) \cdot Y + E(X) \cdot E(Y)$$

Now $E(X)$ and $E(Y)$ are constants, so

$$\begin{aligned} \text{cov}(X,Y) &= E\{[X - E(X)] \cdot [Y - E(Y)]\} \\ &= E(X \cdot Y) - E(Y) \cdot E(X) - E(X) \cdot E(Y) + E(X) \cdot E(Y) \\ &= E(X \cdot Y) - E(X) \cdot E(Y) \end{aligned}$$

This completes the derivation.

The multiplication rule for expected values (page 118) has the following corollary:

(3) If X and Y are independent, then $\operatorname{cov}(X,Y) = 0$

The converse of (3) may be false: there are exceptional cases where $\operatorname{cov}(X,Y) = 0$, but X and Y are dependent. See exercise 8 below.

(4) Proposition. Let X and Y be random variables, dependent or independent. Let a, b, c be constants. Then

$$\operatorname{var}(aX + bY + c) = a^2 \operatorname{var} X + 2ab \operatorname{cov}(X,Y) + b^2 \operatorname{var} Y$$

Proof. Let $U = aX + bY + c$. Then

$$E(U) = aE(X) + bE(Y) + c$$

So

$$\begin{aligned} U - E(U) &= [aX + bY + c] - [aE(X) + bE(Y) + c] \\ &= a[X - E(X)] + b[Y - E(Y)] \end{aligned}$$

Therefore

$$[U - E(U)]^2 = a^2[X - E(X)]^2 + 2ab[X - E(X)][Y - E(Y)] + b^2[Y - E(Y)]^2$$

Now

$$\begin{aligned} \operatorname{var} U &= E\{[U - E(U)]^2\} \\ &= a^2E\{[X-E(X)]^2\} + 2abE\{[X-E(X)][Y-E(Y)]\} + b^2E\{[Y-E(Y)]^2\} \\ &= a^2 \operatorname{var} X + 2ab \operatorname{cov}(X,Y) + b^2 \operatorname{var} Y \end{aligned}$$

This completes the proof.

(5) <u>Corollary</u>. Let X and Y be independent random variables. Let a, b, c be constants. Then

$$\text{var}(aX + bY + c) = a^2 \cdot \text{var } X + b^2 \cdot \text{var } Y$$

<u>Proof</u>. Substitute (3) into (4).

This corollary justifies the addition rule for variances (page 123).

<u>Example</u>. Define X and Y as follows:

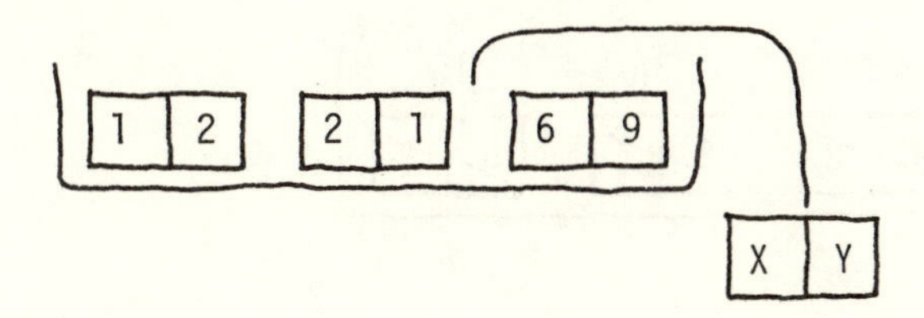

(a) Find $\text{cov}(X,Y)$

(b) Find $\text{var}(X + Y)$ by direct calculation

(c) Find $\text{var } X$ and $\text{var } Y$

(d) Verify that $\text{var}(X + Y) = \text{var } X + 2\text{cov}(X,Y) + \text{var } Y$

<u>Solution</u>. First,

$$E(X) = \frac{1 + 2 + 6}{3} = 3 \quad \text{and} \quad E(Y) = \frac{2 + 1 + 9}{3} = 4$$

For part (a), write the value of $[X - E(X)] \cdot [Y - E(Y)]$ on each ticket:

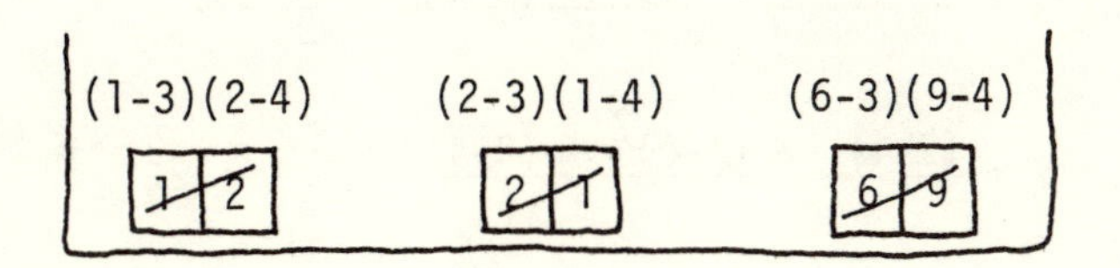

Now

$$\text{cov}\,(X,Y) = E\{[X - E(X)]\cdot[Y - E(Y)]\}$$

$$= \frac{(1-3)(2-4) + (2-3)(1-4) + (6-3)(9-4)}{3}$$

$$= \frac{4 + 3 + 15}{3}$$

$$= \frac{22}{3} = 7\frac{1}{3}$$

For part (b), write the value of $X + Y$ on each ticket:

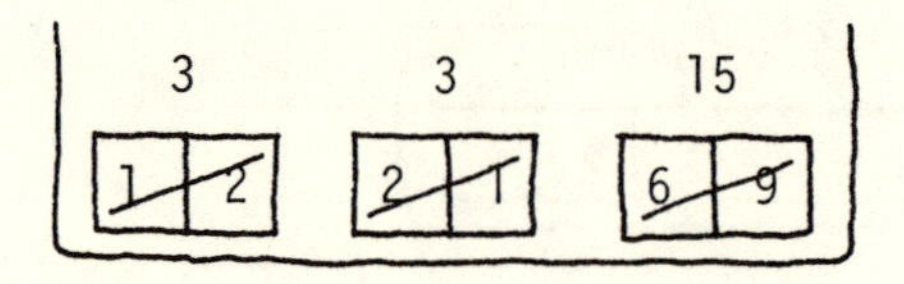

Now

$$E(X + Y) = \frac{3 + 3 + 15}{3} = 7$$

$$\text{var}\,(X + Y) = \frac{(3-7)^2 + (3-7)^2 + (15-7)^2}{3}$$

$$= \frac{16 + 16 + 64}{3}$$

$$= \frac{96}{3} = 32$$

For part (c),

$$\text{var}\,X = \frac{(1-3)^2 + (2-3)^2 + (6-3)^2}{3} = \frac{14}{3} = 4\frac{2}{3}$$

$$\text{var}\,Y = \frac{(2-4)^2 + (1-4)^2 + (9-4)^2}{3} = \frac{38}{3} = 12\frac{2}{3}$$

For part (d), the equality holds, because

$$32 = 4\frac{2}{3} + 2\cdot 7\frac{1}{3} + 12\frac{2}{3}$$

<u>Exercise set 26</u>

(This exercise set covers material from pervious sections.)

1. Show that $E\{X - E(X)\} = 0$.

   Hint: Use the addition rule on page 116.

2. Show that $\text{var } X = \text{cov}(X,X)$.

   Hint: Put $Y = X$ in formula (1).

3. Show that $\text{var } X = E(X^2) - [E(X)]^2$.

   Hint: Put $Y = X$ in formula (2).

4. Show that $\text{cov}(X,Y) = \text{cov}(Y,X)$

   Hint: $XY = YX$.

5. Show that $\text{cov}(X + Y,Z) = \text{cov}(X,Z) + \text{cov}(Y,Z)$.

   Hint: $(X + Y)Z = XZ + YZ$.

6. Show that $\text{cov}(aX + bY,Z) = a\cdot\text{cov}(X,Z) + b\cdot\text{cov}(Y,Z)$.

   Hint: $(aX + bY)Z = aXZ + bYZ$.

7. In the example on page 133, are X and Y independent?

8. Repeat the example on page 133, with X and Y defined as follows:

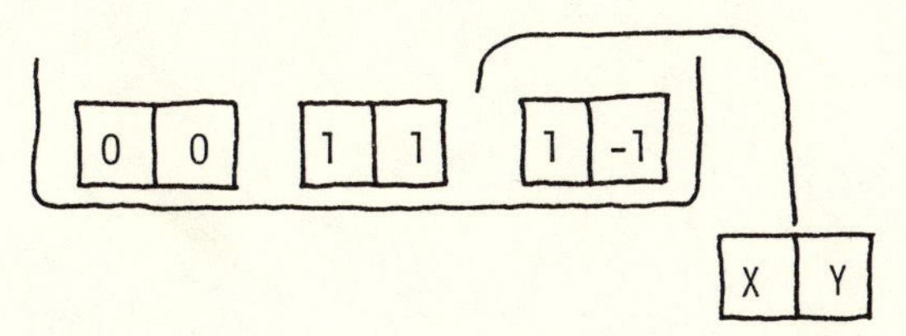

   True or false, and explain: with this box, X and Y are dependent but $\text{cov}(X,Y) = 0$.

PART C. INFERENCE

27. Box models[1]

Many chance quantities are like draws from a box, so random variables can be used to analyze chance variability in many different situations. Here are some examples.

Example 1. A die is rolled 100 times. The total number of spots will be around ______ , give or take ________ or so.

Solution. The key step is to make a box model. Rolling a die and counting the number of spots is like drawing a ticket at random from the box

Why? For instance, the number of spots will be 1 with one chance in six; so will a draw from the box. Similarly, the number of spots will be 2 with one chance in six, and so will a draw from the box. The same applies to the other possible values, 3, 4, 5, and 6. In other words, as far as values and chances go, there is no difference between a draw from the box, and a roll of the die. So much for one roll. Making a hundred rolls of the die is like drawing a hundred times at random with replacement from the box. Why "with replacement"? Because the rolls of the die are independent; and so are the draws from the box.

The problem is about the total number of spots. To model this, we just add up the draws from the box. In short, the total number of spots is like the sum of a hundred draws from the box. Write X_1 for the 1st draw, X_2 for the 2nd,..., and X_{100} for the 100th. The total number of spots is

---

[1] Text, chapter 16.

like the random variable T, where

$$T = X_1 + X_2 + \dots + X_{100}$$

Here, "is like" has a very precise sense: the possible values and the chances are the same. Now the expected value and standard error come into play. See example 25.4 on page 126: the number of spots will be around 350, give or take 17 or so. This completes the solution.

The next example is about roulette.[1] A Nevada roulette wheel has 38 pockets: one is numbered 0, one is numbered 00, and the rest are numbered from 1 through 36. The wheel is spun, the croupier drops the ball, and it lands in each pocket with an equal chance, 1 in 38. One bet is "even-or-odd". Put a dollar bet on "evens": you win a dollar if the ball lands in one of the 18 even pockets numbered 2, 4,..., 36; lose a dollar if the ball lands in one of the 18 odd pockets numbered 1, 3,..., 35; and also lose a dollar if the ball lands in one of the 2 special pockets 0 and 00. These give the house its edge. There are 18 chances in 38 to win a dollar, and 20 chances in 38 to lose a dollar. The amount you win or lose is your <u>net gain</u>.

<u>Example 2</u>. You play roulette 100 times, putting a dollar on evens each time. Your net gain will be around ________, give or take ________ or so.

<u>Solution</u>. Your net gain is like the sum of 100 draws made at random with replacement from the following box:

18 +\$1 20 -\$1

[1]Text, pp. 248-251.

The values on the tickets show how much you win or lose on a single play; the numbers of tickets are adjusted to make the chances right. Now

$$\text{average of box} = \frac{18 \cdot 1 + 20 \cdot (-1)}{38} = -\frac{1}{19} \approx \$0.05$$

By formula 7.3 on page 34,

$$\text{variance of box} = \frac{18 \cdot 1^2 + 20 \cdot (-1)^2}{38} - (-\frac{1}{19})^2 \approx 1$$

$$\text{SD of box} \approx \$1$$

As in example 24.2 on page 117,

$$\text{expected value of sum} = 100 \cdot \text{average of box} \approx -\$5$$

By the square root law (page 125),

$$\text{SE of sum} = \sqrt{100} \cdot \text{SD of box} \approx 10 \cdot \$1 = \$10$$

You will lose around \$5, give or take \$10 or so. This completes the solution.

The next examples are about algebraic operations on random variables.

Example 3. The roulette wheel is spinning. Gambler A bets \$1 on evens; let X stand for his net gain. Gambler B bets \$5 on evens; let Y stand for her net gain. Express Y in terms of X.

Solution. The answer is that $Y = 5X$. If an even number comes up, then A wins \$1 and B wins \$5; if an odd number comes up, or 0, or 00, then A loses \$1 and B loses \$5. Either way,

$$\text{B's net gain} = 5 \cdot (\text{A's net gain})$$

$$Y = 5 \cdot X$$

Another bet in roulette is on a single number. Suppose you bet a dollar on 17. If the ball lands in that pocket, you win \$35; if not, you lose the dollar. Of course, you can spread your bets, for instance by putting a dollar on 17 and a dollar on evens. The house settles each bet separately.

Example 4. The roulette wheel is spinning. Gambler A bets \$1 on evens; let X stand for his net gain. Gambler B bets \$1 on the single number 17; let Y stand for his net gain. Gambler C bets \$5 on evens and \$10 on 17; let Z stand for her net gain. Express Z in terms of X and Y. Find $E(Z)$.

Solution. First, $Z = 5X + 10Y$: the reasoning is as in the previous example. Then $E(Z) = 5E(X) + 10E(Y)$ by the addition rule (page 116). But

$$E(X) = -\$1/19 \quad \text{and} \quad E(Y) = -\$1/19$$

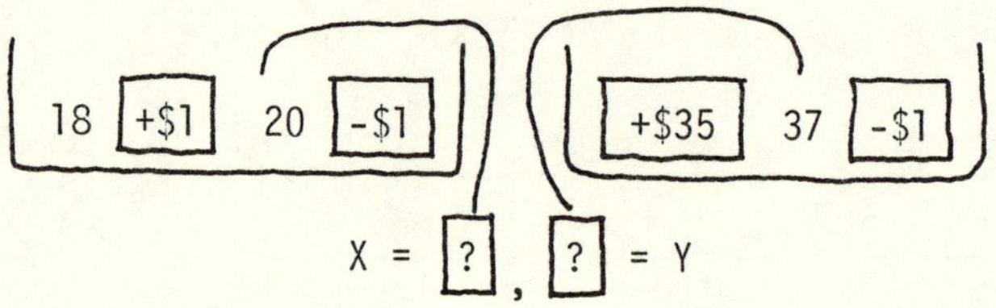

18 [+\$1] 20 [-\$1]   [+\$35] 37 [-\$1]

X = [?] , [?] = Y

So $E(Z) = -\$15/19 \approx -\$0.79$, completing the solution. Notice that gambler C staked \$15. In general, the expected net gain is negative, and 1/19 of the stake. The 1/19 is the house cut.

Exercise set 27

1. Suppose you play roulette 1000 times, staking \$1 on the number 17 each time. Your net gain will be around ______, give or take ______ or so. Explain.

2. Gambler A bets \$18 on evens, playing only once. Gambler B plays 18 times, betting \$1 on evens each time. Is there any difference between these two systems? Explain.

3. The roulette wheel is spinning. Gambler A bets \$1 on each of the 18 even numbers 2, 4,..., 36. Gambler B bets \$18 on evens. Is there any difference between these two systems? Explain.

4. (a) The roulette wheel is spinning. You bet \$5 on the number 17, \$10 on the number 5, and \$25 on the number 12. Your total stake is \$5 + \$10 + \$25 = \$40. What is the expected value of your net gain? Explain.

   (b) The roulette wheel is spinning. You bet $\$s_1$ on the number n_1, and $\$s_2$ on the number n_2, and $\$s_3$ on the number n_3. Your total stake is $\$s = \$s_1 + \$s_2 + \$s_3$. What is the expected value of your net gain?

   Hint: The notation in part (b) may be a bit intimidating. To unscramble it: part (a) is a special case of (b), with $s_1 = 5$ and $n_1 = 17$, $s_2 = 10$ and $n_2 = 5$, $s_3 = 25$ and $n_3 = 12$.

5. Gambler A plays roulette 10 times, betting \$1 on the number 17 each time. Let her net gain on the 1st play be X_1, her net gain on the 2nd play be $X_2, \ldots,$ her net gain on the 10th play be X_{10}. On the first play, gambler B bets \$10 on the number 17, and then he stops betting. On the first five plays, gambler C bets \$1 on 17, and then he stops betting. Match the formula with the phrase, and explain:

[Continues on next page.]

$10X_1$	net gain of gambler A
$X_1+\ldots+X_5$	net gain of gambler B
$X_1+\ldots+X_{10}$	net gain of gambler C

6. The roulette wheel is spinning. Gambler A bets $1 on evens; let X stand for his net gain. Gambler B bets $1 on odds; let Y stand for her net gain. True or false, and explain:

   (a) X and Y have the same distribution

   (b) X and Y are always equal to each other

   (c) Y can be expressed in terms of X

   (d) X and Y are independent

28. Inference[1]

So far, we have been considering situations where the contents of the box are known, and the problem is to find the chances for the draws. There is a converse problem, where the contents of the box are unknown. We make some draws, and try to reason from the results of the draws to the contents of the box. This is called inference.

Ordinarily, the draws are made without replacement. However, in practice there is often very little difference between drawing with or without replacement, because the sample is such a small part of the population. For example, the Gallup Poll predicts election results using a sample of about 5,000 voters out of 100,000,000. In this section, we will assume the draws are made with replacement. This keeps the mathematics from getting too complicated, and often provides a good approximation for draws made without replacement.[2]

It will help to begin with a case study. A Group Dental Practice in California was engaged in litigation.[3] At the time of the study, 53,872 clients were enrolled in the practice; each had a chart, showing the total amount of money billed for dental work on the client by the practice. These charts were numbered consecutively from 1 on up: the first client to enroll got #1, the second got #2, and so on.

To assess damages in the litigation, it was desired to know the average billing for all 53,872 clients. In principle, this could have been done by reviewing all the charts, but such a review would have been too expensive. So 400 charts were drawn at random (for purposes of this section, assume the draws were made with replacement). The total billing was determined for each chart

---

[1]Supplements part VI of the text.

[2]Text, page 329.

[3]The parties to this litigation cannot be identified. Some of the numbers have been changed.

in the sample. The average of these 400 numbers was \$569. However, there was a lot of spread: the standard deviation of these 400 numbers was \$733. The statistical problem was to estimate the average billing for all 53,872 clients, and put a give-or-take number on the estimate.

A box model for this problem can be defined as follows: n draws are made at random with replacement from a box, whose contents are unknown. In the case study, there were 53,872 tickets in the box, one for each client; the total billing for the client is shown on the corresponding ticket: and n is 400. Naturally the average of the draws is used to estimate the average of the box. However, there is some error in this estimate. How much? The exact size of the error cannot be determined, unless the average of the box is known. However, the likely size of the error can be determined, and this is often good enough.

Let $y_1, y_2, \ldots, y_n$ denote the numbers drawn from the box -- the observed values. In the case study, y_1 is the total billing for the first client in the sample. Likewise for y_2, and so on. The average of the box is estimated as

(1) $$\bar{y} = \frac{1}{n} \sum_{i=1}^{n} y_i$$

Let $Y_1, Y_2, \ldots, Y_n$ denote the draws from the box -- the random variables. Then $\bar{y}$ is the observed value of a random variable $\bar{Y}$:

(2) $$\bar{Y} = \frac{1}{n} \sum_{i=1}^{n} Y_i$$

Let μ be the average of the box, and let σ^2 be the variance of the box. Here, μ and σ^2 are parameters, that is, characteristics of the model.[1]

[1]The number of tickets in the box is a parameter. So is the size of the sample. The value of a parameter may be known, or unknown. The greek letter μ is read "mu"; and σ as "sigma".

They are not random variables, or observed values, but unknown numbers. The average of the box is not changing, and it was not determined by a chance procedure: it is just a number we want to estimate.[1] The observed value of $\bar{Y}$ is used to estimate μ, and the question is, how far off is the estimate likely to be?

$\bar{Y}$ is around μ, give or take ___?___ or so

It is helpful to make two computations from the point of view of an omniscient statistician, who knows μ and σ^2; then we have to connect his results with our problem. First, by the addition rule for expected values (page 116),

$$E(\bar{Y}) = E\{\frac{1}{n}(Y_1 + Y_2 + \ldots + Y_n)\} = \frac{1}{n}\{E(Y_1) + \ldots + E(Y_n)\}$$

There are n terms in the sum, and each is μ. So

$$E(\bar{Y}) = \mu \tag{3}$$

For the second calculation, the addition rule for variances (page 123) comes into play:

$$\text{var } \bar{Y} = \text{var } \{\frac{1}{n}(Y_1 + Y_2 + \ldots + Y_n)\} = \frac{1}{n^2}\{\text{var } Y_1 + \text{var } Y_2 + \ldots + \text{var } Y_n\}$$

The Y's are independent because the draws are made with replacement. There are n terms in the sum, and each is σ^2. So

$$\text{var } \bar{Y} = \frac{1}{n^2} \cdot n \cdot \sigma^2$$

$$\text{var } \bar{Y} = \sigma^2/n \tag{4}$$

$$\text{SE of } \bar{Y} = \sqrt{\text{var } \bar{Y}} = \sigma/\sqrt{n} \tag{5}$$

---

[1] We are operating within the frequency theory of chance: text, pp. 203 and 347.

So the omniscient statistician knows how well $\bar{Y}$ is likely to do:

$$\bar{Y} \text{ is around } \mu\text{, give or take } (\sigma/\sqrt{n}) \text{ or so}$$

To make use of this result, an ordinary statistician must estimate σ from the data:

$$\bar{Y} \text{ is around } \mu\text{, give or take (estimated } \sigma/\sqrt{n}\text{) or so}$$

How is σ to be estimated? One natural procedure uses the standard deviation of the draws. Let

$$S = \sqrt{\frac{1}{n}\sum_{i=1}^{n}(Y_i - \bar{Y})^2} \qquad (6)$$

So S is the standard deviation of the draws, and S estimates σ. The give-or-take number is $S/\sqrt{n}$, and this completes the solution to the inference problem. In the case study, S turned out to be \$733. So the give-or-take number is $\$733/\sqrt{400} \approx \$37$. The average billing for all 53,872 clients is estimated as \$569, give or take \$37 or so. This completes the example.

In formula (6), $\bar{Y}$ is a random variable: the average of the draws. So is S, the standard deviation of the draws. To be specific, S stands for the following chance procedure: take n draws, compute the standard deviation of the list of n numbers you get. It is the observed value of this random variable which is used to estimate the standard deviation of the box.

There is something quite complicated here. You start with some data, compute the average and SD, divide the SD by $\sqrt{n}$. But in order to make sense out of the computations, it is necessary to think of the data as the list of observed values of some random variables -- the draws from the box. The average of the data has to be considered as the observed value of $\bar{Y}$; the standard deviation of the data, as the observed value of S. That is why inference is a hard topic: it involves the chance procedure which generated the data.

In (6), some statisticians prefer to divide by n-1:

(7) $$(S^+)^2 = \frac{1}{n-1} \sum_{i=1}^{n} (Y_i - \bar{Y})^2$$

They use $(S^+)^2$ to estimate the variance of the box.[1] The argument is that S^2 is a biased estimate of σ^2. (An estimate of a parameter is biased when its expected value differs from the parameter.) In fact, as will be proved below, $E\{S^2\}$ is too small by the factor $(n-1)/n$:

(8) $$E\{S^2\} = \frac{n-1}{n} \sigma^2$$

On the other hand, $(S^+)^2$ is unbiased:

(9) $$E\{(S^+)^2\} = \sigma^2$$

To derive (9) from (8), notice that

$$(S^+)^2 = \frac{n}{n-1} S^2$$

You can check this by comparing formulas (6) and (7). Now

$$E\{(S^+)^2\} = \frac{n}{n-1} E\{S^2\}$$

$$= \frac{n}{n-1} \cdot \frac{n-1}{n} \cdot \sigma^2$$

$$= \sigma^2$$

This completes the derivation of (9) from (8).

To conclude this section, we are going to prove formula (8). This is quite technical, and readers can skip to the exercises without losing their way.

---

[1] Text, pp. 65-66 and pp. 409-415.

The first step is to rewrite (6), using formula 7.3 on page 34:

$$S^2 = \left(\frac{1}{n}\sum_{i=1}^{n} Y_i^2\right) - \bar{Y}^2 \tag{10}$$

Now use the addition rule for expected values (page 116):

$$E(S^2) = \left[\frac{1}{n}\sum_{i=1}^{n} E(Y_i^2)\right] - E(\bar{Y}^2) \tag{11}$$

We have to evaluate all the terms in (11), and start with the last one. By exercise 26.3 on page 135,

$$\operatorname{var} \bar{Y} = E(\bar{Y}^2) - E(\bar{Y})^2$$

Substitute equations (3) and (4) into this:

$$\frac{\sigma^2}{n} = E(\bar{Y}^2) - \mu^2$$

Rearranging,

$$E(\bar{Y}^2) = \mu^2 + \frac{\sigma^2}{n} \tag{12}$$

Similarly,

$$\operatorname{var} Y_i = E(Y_i^2) - E(Y_i)^2$$

So

$$E(Y_i^2) = \operatorname{var} Y_i + E(Y_i)^2$$

$$= \text{var of box} + (\text{ave of box})^2$$

$$= \sigma^2 + \mu^2$$

And

$$\frac{1}{n} \sum_{i=1}^{n} E(Y_i^2) = \sigma^2 + \mu^2$$

Substitute this and (12) into (11):

$$E(S^2) = \sigma^2 + \mu^2 - \mu^2 - \frac{\sigma^2}{n}$$

$$= \frac{n-1}{n} \sigma^2$$

This completes the proof of (8).

Exercise set 28

1. A hundred draws are made at random with replacement from a box of numbered tickets. The average of the list of 100 draws turns out to be 3.7, with an SD of 2.

   (a) The average of the box is around ______, give or take ______ or so.

   (b) Classify each of the following as an observed value or a parameter:

      (i) The average of the list of 100 draws.

      (ii) The average of the box.

      (iii) The SD of the list of 100 draws.

      (iv) The SD of the box.

2. A hundred draws are made at random with replacement from a box of numbered tickets. Let $X_1, X_2, \ldots, X_{100}$ denote the 1st, 2nd,...,100th draw. Match the formula with the phrase. Something may be left over.

[Continues on next page.]

X_7	The average of the 100 draws
$\frac{1}{50}(X_1+X_2+\ldots+X_{50})$	The variance of the 100 draws
$\frac{1}{100}(X_1+X_2+\ldots+X_{100})$	The variance of the average of the 100 draws
$(\frac{1}{100}\sum_{i=1}^{100} X_i^2) - (\frac{1}{100}\sum_{i=1}^{100} X_i)^2$	The average of the first 50 draws

3. Continuing exercise 2, say whether each of the following is true or false, and explain why:

   (a) $X_1+X_2+\ldots+X_{50}$ and $50 \bullet X_1$ are two different formulas for the same random variable

   (b) $\text{var}(X_1+X_2+\ldots+X_{50}) = 50 \bullet \text{var } X_1$

   (c) $\text{var}\{\frac{1}{50}(X_1+X_2+\ldots+X_{50})\} = \frac{1}{50} \bullet \text{var } X_1$

4. Continuing exercise 3, which of the following is smallest? Largest? Explain.

   (i) $\text{var } X_1$

   (ii) $\text{var}(X_1+X_2+\ldots+X_{100})$

   (iii) $\text{var}\{\frac{1}{100}(X_1+X_2+\ldots+X_{100})\}$

5. True or false, and explain: the average of the draws is a single number, so its variance is zero.

6. Suppose $E(\bar{X}) = 3.012$ and $\text{var } \bar{X} = .011$. Is an observed value of 3.121 for $\bar{X}$ surprising?

   Yes ________ No ________

   Explain.

7. Let $X_1, X_2, \ldots, X_n$ be repeated draws made at random with replacement from a box.

(a) Classify each of the following as a random variable or a parameter. Explain carefully.

$E(X_1)$ $\qquad\qquad \bar{X} = \frac{1}{n} \sum_{i=1}^{n} X_i$

$\text{var } X_1$ $\qquad\qquad s^2 = \frac{1}{n} \sum_{i=1}^{n} (X_i - \bar{X})^2$

(b) True or false, and explain:

(i) $E(\bar{X}) = \frac{1}{n} \sum_{i=1}^{n} X_i$

(ii) $\text{var } \bar{X} = \frac{1}{n} \sum_{i=1}^{n} (X_i - \bar{X})^2$

8. Twenty-five draws are made at random with replacement from a box. Classify each of the following as: a parameter, or an observed value, or nonsense.

(i) The variance of the average of the 25 draws.

(ii) The average of the variance of the 25 draws.

(iii) The variance of the 25 draws.

(iv) The average of the 25 draws.

29. The Gauss model[1]

Suppose repeated, independent measurements are made on some quantity. They will differ among themselves. Each is thrown off the exact value by some small error, and this error changes from measurement to measurement. Carl Friedrich Gauss (Germany, 1777-1855) proposed a statistical model to analyze this sort of measurement error. In our terminology, this was a box model for data -- for a list of repeated measurements, like 100 measurements on a check weight.[2]

The model is that each measurement equals the exact value of the quantity being measured, plus an error drawn from a box. The draws are made at random with replacement; this captures the idea of repeated, independent measurements. The numbers in the box represent the possible errors. The average of the box is assumed to be zero: no bias. The standard deviation of the box is a parameter, usually denoted by σ. This parameter indicates the likely size of the chance error, and hence the likely accuracy of any single measurement. Ordinarily, σ is unknown and must be estimated from the data.

The model can be expressed in a diagram, as follows:

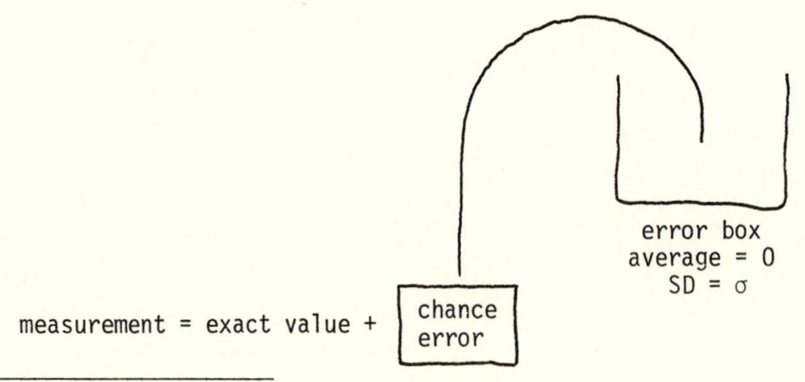

[1]Text, chapter 24.

[2]Text, table 1 on page 91.

As an equation,

$$Y_i = \mu + \varepsilon_i \quad (1)$$

In this equation:

- i indexes the measurements; $i = 1$ means, the 1st measurement; $i = 2$, the 2nd; and so on
- Y_i is the ith measurement
- μ is the exact value; this is a parameter, and it is usually unknown
- ε_i is the ith draw from the error box; this is a random variable, but not directly observable.

To read the equation: μ is the greek letter "mu", and ε is "epsilon". By assumption, $E(\varepsilon_i) = 0$ and $\operatorname{var}\varepsilon_i = \sigma^2$; the random variables $\varepsilon_1, \varepsilon_2, \ldots$ are independent and identically distributed.

In equation (1), the random variable Y_i is observable. Indeed, the model is that the list of repeated measurements is the list of observed values of $Y_1, Y_2, \ldots$. However, the random variable ε_i, the ith draw from the error box, is not directly observable. Only the combination $\mu + \varepsilon_i$ can be observed, and this cannot be split into its two components.

Suppose there are n repeated measurements. Let $y_1, y_2, \ldots, y_n$ be the observed values, and let $\bar{y}$ be the average:

$$\bar{y} = \frac{1}{n} \sum_{i=1}^{n} y_i$$

Then $\bar{y}$ is the estimate for μ. To see how good this estimate is likely to be, we have to consider $\bar{y}$ as the observed value of the random variable $\bar{Y}$:

$$\bar{Y} = \frac{1}{n}(Y_1 + Y_2 + \ldots + Y_n) \quad (2)$$

The Gauss model was designed to answer the question, how close to μ is $\bar{Y}$ likely to be?

$\bar{Y}$ will be around μ, give or take ___?___ or so.

Before answering this question, however, we should consider the deviation of y_i from the average $\bar{y}$. Let e_i be this deviation, so $e_i = y_i - \bar{y}$. Rearranging,

$$y_i = \bar{y} + e_i \tag{3}$$

This looks like equation (1), but it is quite different:

- y_i, $\bar{y}$ and e_i in (3) are numbers, observed values of random variables.
- Y_i in (1) is a random variable, μ is a parameter, and ε_i is an unobservable random variable.

The deviation $e_i = y_i - \bar{y}$ is the observed value of the random variable $Y_i - \bar{Y}$. If n is large, $\bar{Y}$ is close to μ, so $Y_i - \bar{Y}$ is close to ε_i; but there is still going to be some discrepancy. So e_i is not quite the observed value of ε_i, but it is close.

Coming back to the main question, how close is $\bar{Y}$ to μ? As equation (1) shows,

$$\bar{Y} = \mu + \frac{1}{n}(\varepsilon_1 + \varepsilon_2 + \ldots + \varepsilon_n) \tag{4}$$

So the difference between $\bar{Y}$ and μ equals the average of n draws from the error box. By the addition rule for expected values (page 116),

$$E(\bar{Y}) = \mu \tag{5}$$

By the addition rule for variances (page 123),

$$\text{var } \bar{Y} = \frac{1}{n^2}(\text{var } \varepsilon_1 + \text{var } \varepsilon_2 + \ldots + \text{var } \varepsilon_n)$$

$$= \frac{1}{n^2} \cdot n \cdot \sigma^2$$

So

$$\text{var } \bar{Y} = \sigma^2/n \tag{6}$$

(7) $$\text{SE of } \bar{Y} = \sqrt{\text{var } \bar{Y}} = \sigma/\sqrt{n}$$

The argument is only sketched, because the details are like those in the previous section.

Equations (5) and (7) show that:

$\bar{Y}$ will be around μ, give or take $\sigma/\sqrt{n}$ or so

When σ is unknown, it has to be estimated: The ε_i are repeated draws from the error box; the average of this box is zero, so a natural estimate for σ is

$$\sqrt{\frac{1}{n}\sum_{i=1}^{n} \varepsilon_i^2}$$

However, the measurement error ε_i is unobservable: so it too has to be estimated, by the deviation $e_i = y_i - \bar{y}$. Thus, σ can be estimated as

$$\sqrt{\frac{1}{n}\sum_{i=1}^{n} e_i^2}$$

This is just the SD of the data set $y_1, y_2, \ldots, y_n$. (Some statisticians prefer to divide by $n-1$ rather than n, as noted in the previous section.)

This completes the discussion of the technique. When using the formula, however, it is important to remember that the model is entirely hypothetical. So are all the ingredients: μ, σ, and the error box. What investigators have in front of them is the data, the list of measurements. This creates an interesting tension, because to make a sensible analysis of the data, you need the model. That is, an investigator can always compute the SD from the data and divide by $\sqrt{n}$. But if the data show any trend or pattern, the model does not apply: draws from a box do not show trends or patterns. And if the model does not apply, the formulas will give answers which are misleading. The investigator cannot make a valid statistical inference if the model is wrong.[1]

More about the notation: Sometimes, the model is written in quite a brutal way:

(8) $$Y_i = \mu + \varepsilon_i$$

(9) $$E(\varepsilon_i) = 0 \text{ and } \operatorname{var} \varepsilon_i = \sigma^2$$

(10) the ε_i are independent and identically distributed

Let us take this slowly. In equation (9), $E(\varepsilon_i) = 0$ because the average of the error box is 0. Likewise, $\operatorname{var} \varepsilon_i = \sigma^2$ because σ is the SD of the error box. In equation (10), the ε_i are independent because the draws are made with replacement. "Identically distributed" means that all the ε_i have the same distribution. Again, this is so because the draws are made with replacement: ε_{100} comes from the same box as ε_1. The phrase "independent and identically distributed" comes up a lot in statistics, and is sometimes contracted to "IID". So (10) becomes "The ε_i are IID". Some statisticians do not bother writing (9) and (10) at all, but consider them somehow implied by (8). This is just sloppy thinking.

<u>Exercise set 29</u>

1. Listed below are ten repeated, independent measurements on the NB 10 check-weight.[2] Estimate the exact weight, and put a plus-or-minus number on the estimate.

409	400	406	399	402
406	401	403	401	403

[1] Text, pp. 350-352, 366, 404-408.

[2] Text, p. 91.

2. In exercise 1, classify each of the following as an observed value or a parameter:

   (a) the exact weight

   (b) the average of the ten measurements listed above

   (c) the SD of the ten measurements listed above

3. In exercise 1, can you find e_3? ε_3? Explain.

4. In equation (4), classify each of the following as: an observable random variable, an unobservable random variable, a parameter:

$$\bar{Y} \qquad \mu \qquad \varepsilon_1$$

   Explain.

5. True or false, and explain: $\bar{Y}$ is a single number, so $E(\bar{Y}) = \bar{Y}$ and $\text{var}\ \bar{Y} = 0$.

30. The regression model[1]

The regression model is also a box model for data; but this time, the data consist of pairs of numbers, as in table 1 below.

Table 1. Data.

x	y
x_1	y_1
x_2	y_2
.	.
.	.
.	.
x_n	y_n

The regression model can be written as follows:

(1) $$Y_i = \alpha + \beta x_i + \varepsilon_i$$

(2) $$E(\varepsilon_i) = 0 \quad \text{and} \quad \text{var } \varepsilon_i = \sigma^2$$

(3) the ε_i are independent and identically distributed

This is like the Gauss model of the previous section, but it is quite a bit more complicated. The ingredients for (1) are as follows:

- i is an index; $i = 1$ refers to the 1st data point; $i = 2$, to the 2nd; and so on; usually, the number of data points is denoted by n.
- α and β are parameters; usually, these are unknown and have to be estimated from the data.
- x_i is a parameter, but one which is under the control of the investigator.
- ε_i is an unobservable random variable. Sometimes, ε_i is called a "disturbance term".

[1]See chapter 12 of the text; also see sections 11-16 in this booklet.

Assumption (3) says that ε_i are like draws made at random with replacement from a box. Assumption (2) says that the average of the box is 0, and the SD of the box is σ. The σ is a parameter, and usually it too is unknown and has to be estimated from the data.

Assumptions (1), (2) and (3) tell a story about how some data set was generated. The story involves the investigator and nature.

<u>Step 1</u>. The investigator chooses n, the number of data points; the investigator also chooses $x_1, x_2, \ldots, x_n$.

<u>Step 2</u>. Nature makes n draws at random with replacement from the error box, whose average is 0; call these draws $\varepsilon_1, \varepsilon_2, \ldots, \varepsilon_n$.

<u>Step 3</u>. Nature computes $Y_1, Y_2, \ldots, Y_n$ from the formula

$$Y_i = \alpha + \beta x_i + \varepsilon_i$$

The parameters α and β are known to nature, but not to the investigator.

<u>Step 4</u>. The investigator gets to see the values of $Y_1, Y_2, \ldots, Y_n$ but not the ingredients of the model: α, β, the error box, or the ε's.

The observed values of $Y_1, Y_2, \ldots, Y_n$ are the data $y_1, y_2, \ldots, y_n$ in table 1. It is worth noting that the model is completely hypothetical, as are the ingredients: α, β, σ, the error box, the ε's. What the investigator sees is only the data. The statistical object is to estimate α and β, and put give-or-take numbers on the estimates.

At this point, a specific example may be useful. Table 2 reports the results of an experiment on Hooke's law. According to this law, the stretch of a spring is proportional to the load (weight) placed on it. Six different loads were selected by the investigator, and placed in turn on the spring; the loads are shown in the left-hand column of the table. The length of the spring was measured for each load; the lengths are shown in the right-hand column of the table. The data consist of the loads and the lengths -- pairs of numbers. The main object is to estimate the rate at which the spring stretches under load. Let β denote this constant of proportionality; and let α denote the length of the spring under no load.

Table 2. Data on Hooke's law

load (kg)	length (cm)
0	439.00
2	439.12
4	439.21
6	439.31
8	439.40
10	439.50

The index i will be used to refer to the data points in this table: $i = 1$ for the 1st, with a load of 0 and a length of 439.00; $i = 2$ for the 2nd, with a load of 2 and a length of 439.12, and so on. For the ith data point,

$$Y_i = \alpha + \beta x_i + \varepsilon_i$$

where

α = the length of the spring under no load

β = the constant of proportionality

x_i = the load (chosen by the investigator)

$\alpha + \beta x_i$ = the exact length of the spring under the load x_i

ε_i = measurement error

Y_i = measured length of the spring under the load x_i

To repeat, Y_i is a random variable; y_i, its observed value.

In this experiment, length was measured in centimeters: so the units for α, ε_i, and Y_i are cm. Load was measured in kilograms, so the units for x_i are kg. The units for β are cm/kg, stretch per unit of load. The parameters α and β describe mechanical properties of the spring, and are not known to the investigator: the table does not show α, β, or σ. These are all part of the model, and have to be estimated from the data. Nor does the table show the ε_i: these measurement errors are not directly observable.

The parameters α and β are estimated by "the method of least squares", namely by the $\hat{\alpha}$ and $\hat{\beta}$ which minimize

$$\sum_{i=1}^{n} (y_i - \hat{\alpha} - \hat{\beta}x_i)^2$$

(It is a convention of statistics to denote estimates by hats.) In short, $\hat{\beta}$ is the slope of the regression line of y on x, and $\hat{\alpha}$ is the intercept, as discussed in section 11 of this booklet. The next example gives a brief review of some of the material in section 11. It shows how to estimate the parameters α and β in the regression model, but not how to put give-or-take numbers on these estimates.

Example 1. Estimate the parameters α and β from the data in table 2.

Solution. We compute[1]

ave load = 5 kg SD = 3.415650 kg

ave length = 439.256667 cm SD = .167995 cm

r = .999167

[1]The arithmetic is quite annoying. It is necessary to keep many decimals in the intermediate steps in order to get accuracy to (say) three decimals in the final answers.

Now x = load, y = length,

(4) $$\hat{\beta} = r \cdot \text{SD of } y/\text{SD of } x$$

(5) $$\hat{\alpha} = \bar{y} - \hat{\beta}\bar{x}$$

Substituting,

$$\hat{\beta} = \text{slope} = .999167 \times .167995/3.415650 \approx .049 \text{ cm/kg}$$

$$\hat{\alpha} = \text{intercept} = 439.256667 - .049 \times 5 \approx 439.011 \text{ cm}$$

See equations 11.2 and 11.3 on page 55. This completes the solution: the length α of the spring under no load is estimated as 439.011 cm; the spring stretches at the rate β estimated as .049 cm/kg.

The next main task is to put give-or-take numbers on these estimates, but this involves quite a bit of work. To begin with, the regression line leads to the following equation:

(6) $$y_i = \hat{\alpha} + \hat{\beta}x_i + e_i$$

This equation looks like (1), but it is quite different. In (6),

- y_i is a number
- $\hat{\alpha}$ and $\hat{\beta}$ are numbers, namely the intercept and slope computed from the data
- e_i is a number, namely, $y_i - \hat{\alpha} - \hat{\beta}x_i$

This e_i is called a residual. It is important to distinguish between the residual e_i in equation (6), and the disturbance term ε_i in equation (1):

$$e_i = y_i - \hat{\alpha} - \hat{\beta}x_i$$

$$\varepsilon_i = Y_i - \alpha - \beta x_i$$

So e_i is a number, and ε_i is a random variable. Furthermore, e_i involves the estimates for the parameters, and ε_i the parameters themselves. So e_i

is not an observed value of ε_i: indeed, the ε_i are unobservable. If n is large, then the estimates for the parameters will be quite close, so e_i is a good estimate for what ε_i must have been, but there is still some difference: see figure 1.

Figure 1. Residuals and disturbance terms. The solid line is $y = \alpha + \beta x$. This is not observable. The disturbances ε_i are relative to this line, and are not observable either. The dashed line is the regression line $y = \hat{\alpha} + \hat{\beta}x$. This is computed from the data. The residuals e_i are relative to this line. So e_i can be computed from the data.

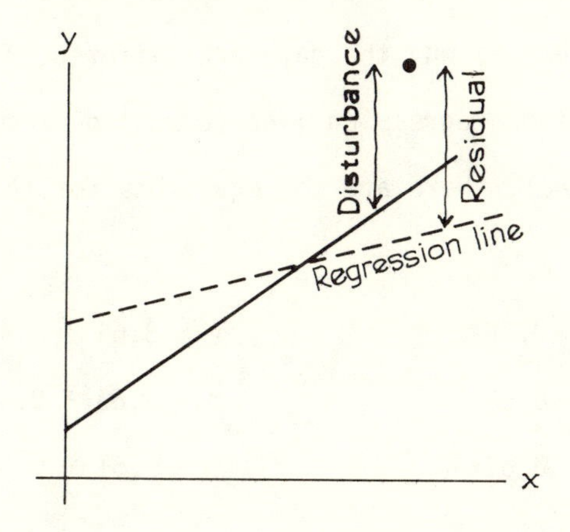

How good are the estimates $\hat{\alpha}$ and $\hat{\beta}$? Before answering this question, we have to see why the estimates are off the parameters in the first place. The reason is the disturbance terms ε_i in equation (1). If these terms were all zero, then the data would fall exactly on the line $\alpha + \beta x$, and the estimates would be exactly right. However, the ε's are not zero, so the points do not fall on the line. Furthermore, the ε's are variable: if an investigator did the experiment again, she would get new values for the ε's, new data points, and a new regression line. There is chance error in $\hat{\alpha}$ and $\hat{\beta}$.

This idea is illustrated in figure 2, which shows six data sets generated from the same regression model. In all six panels,

$$Y_i = \alpha + \beta x_i + \varepsilon_i$$

The parameters in all six panels were $\alpha = 3.00$ and $\beta = 0.50$; the solid line is

$$y = 3.00 + 0.50x$$

But each time the computer made one of the panels, it selected new values for the ε's, at random. That is why the data are different from panel to panel. That is also why the dashed regression line (computed from the data) bounces around from panel to panel. Here are the equations for the six dashed lines:

$y = 3.30 + 0.46x$	$y = 3.64 + 0.41x$
$y = 2.91 + 0.52x$	$y = 3.68 + 0.47x$
$y = 1.43 + 0.61x$	$y = 2.61 + 0.53x$

Figure 2. Chance variability in the regression line (dashed).

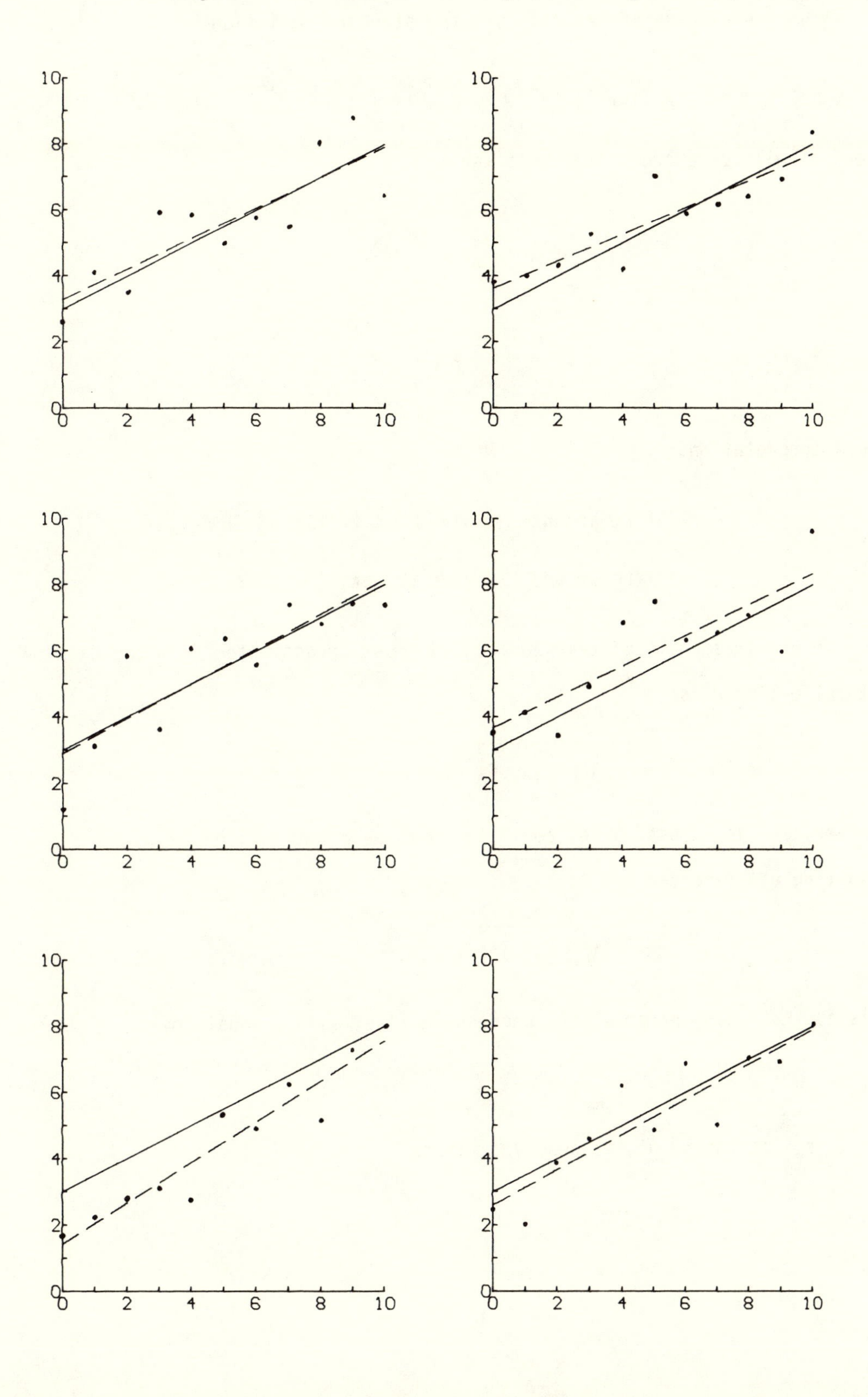

The accuracy question will now be answered, by computing the standard errors. Let $\bar{x}$ be the average of x, and s_x the standard deviation:

$$(7) \qquad \bar{x} = \frac{1}{n}\sum_{i=1}^{n} x_i \quad \text{and} \quad s_x^2 = \frac{1}{n}\sum_{i=1}^{n} (x_i - \bar{x})^2$$

Then, as will be shown in the next section,

$$(8) \qquad \text{SE of } \hat{\alpha} = \frac{\sigma}{\sqrt{n}} \cdot \sqrt{1 + \frac{\bar{x}^2}{s_x^2}}$$

$$(9) \qquad \text{SE of } \hat{\beta} = \frac{\sigma}{\sqrt{n}} \cdot \frac{1}{s_x}$$

The interpretation:

$\hat{\alpha}$ will be around α, give or take (SE of $\hat{\alpha}$) or so

$\hat{\beta}$ will be around β, give or take (SE of $\hat{\beta}$) or so

To use these give-or-take numbers, σ must be estimated from the data. A natural estimate is

$$\sqrt{\frac{1}{n}\sum_{i=1}^{n} \varepsilon_i^2}$$

However, ε_i is unobservable, so it too must be estimated, by e_i. The resulting estimate for σ is

$$(10) \qquad \sqrt{\frac{1}{n}\sum_{i=1}^{n} e_i^2}$$

This is the r.m.s. error of the regression line, and is equal to

$$\sqrt{1 - r^2} \cdot s_y$$

Here, s_y is the SD of the observed values y_i:

$$(11) \qquad \bar{y} = \sum_{i=1}^{n} y_i \quad \text{and} \quad s_y^2 = \frac{1}{n}\sum_{i=1}^{n} (y_i - y)^2$$

Many statisticians prefer to divide by n-2 in (10), rather than n. This will be discussed in the next section, but the reason is bias, as discussed on page 149 above. The official estimate for σ is

$$\hat{\sigma} = \sqrt{\frac{1}{n-2}\sum_{i=1}^{n} e_i^2}$$

That is,

$$\hat{\sigma} = \sqrt{\frac{n}{n-2}} \cdot \sqrt{\frac{1}{n}\sum_{i=1}^{n} e_i^2}$$

Or

$$(12) \qquad \hat{\sigma} = \sqrt{\frac{n}{n-2}} \cdot \sqrt{1 - r^2} \cdot s_y$$

Example 2. Put give-or-take numbers on the estimates $\hat{\alpha}$ and $\hat{\beta}$ in example 1.

Solution. Here n = 6. Substituting into (12),

$$\hat{\sigma} = \sqrt{\frac{6}{4}} \times \sqrt{1 - .999167^2} \times .167995$$

$$\approx .0084$$

The SD of the error box is estimated as .0084 cm: this is the estimate of the typical size of the ε's in equation (1). Now (8) can be used to estimate the SE of $\hat{\alpha}$:

$$\frac{.0084}{\sqrt{6}} \times \sqrt{1 + \frac{5^2}{3.415650^2}} \approx .006 \text{ cm}$$

Thus, α is estimated as 439.011 cm, give or take .006 cm or so.

Likewise, (9) can be used to estimate the SE of $\hat{\beta}$:

$$\frac{.0084}{\sqrt{6}} \times \frac{1}{3.415650} \approx .001 \text{ cm/kg}$$

Thus, β is estimated as .049 cm/kg, give or take .001 cm/kg or so.

Exercise set 30

1. For the data set in table 2 on page 162: find e_3 and ε_3, if possible.

2. Classify the entries in table 2 as: random variables, observed values, parameters.

3. Estimate the unloaded length of the spring and the constant of proportionality from the data in table 1 on page 59. Classify these estimates as: random variables, observed values, parameters. Put plus-or-minus numbers on the estimates.

4. In equation (1), find $E(Y_i)$.

5. Is $E(Y_i)$ the same as $\bar{Y}$? Explain carefully.

6. In an experiment on Hooke's law, the following results are obtained, with 25 data points, x = load, y = length:

$$\sum_{i=1}^{25} x_i = 150 \text{ kg} \qquad \sum_{i=1}^{25} x_i^2 = 1212 \text{ kg}^2$$

$$\sum_{i=1}^{25} y_i = 6428 \text{ cm} \qquad \sum_{i=1}^{25} y_i^2 = 1{,}652{,}812 \text{ cm}^2$$

$$\sum_{i=1}^{25} x_i y_i = 38{,}686 \text{ kg} \times \text{cm}$$

[Continues on next page.]

(a) Estimate the length of the spring under no load, and the constant of proportionality in Hooke's law.

(b) Put plus-or-minus numbers on these estimates.

(c) Classify the following as: random variables, observed values, parameters.

The length of the spring under no load.

The constant of proportionality in Hooke's law.

7. In equation (8), classify each of the following as: an observable random variable, an unobservable random variable, an observed value, a known parameter, an unknown parameter:

$$\hat{\alpha} \qquad \sigma \qquad n \qquad \bar{x} \qquad s_x$$

8. It will be shown in the next section that

$$\hat{\beta} - \beta = \frac{1}{n} \sum_{i=1}^{n} (x_i - \bar{x})\varepsilon_i / s_x^2$$

(a) What does n represent? i? $\bar{x}$? s_x? ε_i?

(b) Classify each of the following as: an observable random variable, an unobservable random variable, an observed value, a known parameter, an unknown parameter:

$$\hat{\beta} \qquad \beta \qquad \bar{x} \qquad \varepsilon_i$$

31. More on the regression model[1]

The object in this section is to prove formulas 30.8 and 30.9 for the standard errors of $\hat{\alpha}$ and $\hat{\beta}$. These standard errors indicate the likely size of the chance errors in the estimates. To derive the formulas, it is necessary to think about random variables rather than observed values. That is, $\hat{\alpha}$ and $\hat{\beta}$ must be visualized as the result of computational procedures based on draws from the error box. In other words, we have to express $\hat{\alpha}$ and $\hat{\beta}$ in terms of the ε's. This takes a bit of work, and 30.4 is not a convenient formula to use. It is better to start this way:

$$\hat{\beta} = [\tfrac{1}{n} \textstyle\sum_{i=1}^{n} (x_i - \bar{x})y_i]/s_x^2 \tag{1}$$

In this formula, $\bar{x}$ is the average of the x's, and s_x is the standard deviation. See exercise 11.5 on page 60.

Remember that the y_i's are the observed values of the Y_i's. So, our estimated slope is really the observed value of the following random variable, which will be denoted by $\hat{\beta}$ too:

$$\hat{\beta} = [\tfrac{1}{n} \textstyle\sum_{i=1}^{n} (x_i - \bar{x})Y_i]/s_x^2 \tag{2}$$

Now substitute formula 30.1 for Y_i into the numerator of this formula, and bring the constants out: the numerator is

$$\frac{1}{n} \sum_{i=1}^{n} (x_i - \bar{x})Y_i$$

This equals

$$\alpha \cdot \frac{1}{n} \sum_{i=1}^{n} (x_i - \bar{x}) + \beta \cdot \frac{1}{n} \sum_{i=1}^{n} (x_i - \bar{x})x_i + \frac{1}{n} \sum_{i=1}^{n} (x_i - \bar{x})\varepsilon_i$$

[1] This section is quite technical, and can be skipped.

But (proposition 5.8 on page 20)

$$\frac{1}{n} \sum_{i=1}^{n} (x_i - \bar{x}) = 0$$

And (formula 7.4 on page 34)

$$\frac{1}{n} \sum_{i=1}^{n} (x_i - \bar{x})x_i = \frac{1}{n} \sum_{i=1}^{n} (x_i - \bar{x})^2 = s_x^2$$

So

$$\frac{1}{n} \sum_{i=1}^{n} (x_i - \bar{x})Y_i = \beta \cdot s_x^2 + \frac{1}{n} \sum_{i=1}^{n} (x_i - \bar{x})\varepsilon_i$$

Substitute back into (2):

(3) $$\hat{\beta} = \beta + \frac{1}{n} \sum_{i=1}^{n} (x_i - \bar{x})\varepsilon_i / s_x^2$$

Formula (3) expresses $\hat{\beta}$ in terms of:

β, the unknown parameter being estimated

x_i, the parameters controlled by the investigator

ε_i, the disturbance terms

The chance error in $\hat{\beta}$ is

(4) $$\hat{\beta} - \beta = \frac{1}{n} \sum_{i=1}^{n} (x_i - \bar{x})\varepsilon_i / s_x^2$$

We can now show that $\hat{\beta}$ is unbiased.

(5) Lemma. $E(\hat{\beta}) = \beta$

Proof. First, $E(\varepsilon_i) = 0$ by 30.2; then apply the addition rule for expected values (page 116) to formula (3). This completes the proof.

Remember that SE of $\hat{\beta} = \sqrt{\text{var } \hat{\beta}}$. So we can prove 30.9 by computing var $\hat{\beta}$.

(6) Lemma. $\text{var } \hat{\beta} = \frac{\sigma^2}{n} \cdot \frac{1}{s_x^2}$

Proof. The ε_i are independent, by 30.3. Now apply the addition rule for variances (page 123) to formula (3):

$$\text{var } \hat{\beta} = \frac{1}{n^2} \sum_{i=1}^{n} (x_i - \bar{x})^2 \cdot \text{var } \varepsilon_i \cdot \frac{1}{s_x^4}$$

But $\text{var } \varepsilon_i = \sigma^2$ for all i, by 30.2. And by 30.7,

$$\sum_{i=1}^{n} (x_i - \bar{x})^2 = n \cdot s_x^2$$

Therefore

$$\text{var } \hat{\beta} = \frac{1}{n^2} \cdot n \cdot s_x^2 \cdot \sigma^2 \cdot \frac{1}{s_x^4}$$

$$= \frac{\sigma^2}{n} \cdot \frac{1}{s_x^2}$$

This completes the argument.

We now have to look at $\hat{\alpha}$. In terms of the data, by formula 30.5,

$$\hat{\alpha} = \bar{y} - \hat{\beta}\bar{x}, \quad \text{where } \bar{y} = \sum_{i=1}^{n} y_i$$

So, our estimated intercept is the observed value of the following random variable, which will be denoted by $\hat{\alpha}$ too:

(7) $$\hat{\alpha} = \bar{Y} - \hat{\beta}\bar{x}, \quad \text{where } \bar{Y} = \frac{1}{n} \sum_{i=1}^{n} Y_i$$

In view of 30.1,

$$\bar{Y} = \alpha + \beta\bar{x} + \frac{1}{n}\sum_{i=1}^{n} \varepsilon_i$$

Substitute into (7):

$$\hat{\alpha} = \alpha - (\hat{\beta} - \beta)\bar{x} + \frac{1}{n}\sum_{i=1}^{n} \varepsilon_i \tag{8}$$

The chance error in $\hat{\alpha}$ is

$$\hat{\alpha} - \alpha = -(\hat{\beta} - \beta)\bar{x} + \frac{1}{n}\sum_{i=1}^{n} \varepsilon_i$$

This involves $\hat{\beta} - \beta$, the chance error in $\hat{\beta}$; this quantity too can be expressed in terms of the ε's: see formula (4).

We can now show that $\hat{\alpha}$ is unbiased.

(9) Lemma. $E(\hat{\alpha}) = \alpha$

Proof. Formula 30.2 gives $E(\varepsilon_i) = 0$. In view of (5),

$$E(\hat{\beta} - \beta) = E(\hat{\beta}) - \beta = 0$$

Now we use (8):

$$E(\hat{\alpha}) = \alpha - E(\hat{\beta} - \beta)\bar{x} + \frac{1}{n}\sum_{i=1}^{n} E(\varepsilon_i) = \alpha$$

This completes the proof.

The next lemma will help in computing $\operatorname{var} \hat{\alpha}$.

(10) Lemma. $\operatorname{cov}(\varepsilon_j, \hat{\beta} - \beta) = \frac{1}{n}(x_j - \bar{x}) \cdot \sigma^2/s_x^2$

Proof. As in exercise 26.6 on page 135,

$$\text{(11)} \qquad \text{cov}\,(U, \textstyle\sum_{j=1}^{n} c_j V_j) = \textstyle\sum_{j=1}^{n} c_j \,\text{cov}\,(U, V_j)$$

Use (4):

$$\text{(12)} \qquad \text{cov}\,(\varepsilon_j, \hat{\beta} - \beta) = \frac{1}{n} \textstyle\sum_{i=1}^{n} (x_i - \bar{x})\,\text{cov}\,(\varepsilon_j, \varepsilon_i)/s_x^2$$

If $i \neq j$, then $\text{cov}\,(\varepsilon_j, \varepsilon_i) = 0$ because ε_j and ε_i are independent (formula 26.3 on page 132). If $i = j$, then

$$\text{cov}\,(\varepsilon_j, \varepsilon_i) = \text{cov}\,(\varepsilon_j, \varepsilon_j) = \text{var}\,\varepsilon_j = \sigma^2$$

This uses exercise 26.2 on page 135. So

$$\text{(13)} \qquad \text{cov}\,(\varepsilon_j, \hat{\beta} - \beta) = \frac{1}{n}(x_j - \bar{x}) \cdot \sigma^2/s_x^2$$

This completes the argument.

(14) Lemma. $\text{cov}\,(\frac{1}{n} \sum_{j=1}^{n} \varepsilon_j, \hat{\beta} - \beta) = 0$

Proof. As in (11),

$$\text{cov}\,(\textstyle\sum_{j=1}^{n} c_j U_j, V) = \textstyle\sum_{j=1}^{n} c_j \,\text{cov}\,(U_j, V)$$

so

$$\begin{aligned}\text{cov}\,(\tfrac{1}{n} \textstyle\sum_{j=1}^{n} \varepsilon_j, \hat{\beta} - \beta) &= \frac{1}{n} \textstyle\sum_{j=1}^{n} \text{cov}\,(\varepsilon_j, \hat{\beta} - \beta) \\ &= \frac{1}{n^2} \textstyle\sum_{j=1}^{n} (x_j - \bar{x}) \cdot \sigma^2/s_x^2 \\ &= 0\end{aligned}$$

This completes the proof.

We can now justify 30.8, by computing var $\hat{\alpha}$: remember that SE of $\hat{\alpha} = \sqrt{\text{var } \hat{\alpha}}$.

(15) Lemma. var $\hat{\alpha} = \frac{\sigma^2}{n} \cdot \{1 + \frac{\bar{x}^2}{s_x^2}\}$

Proof. Apply proposition 26.4 on page 132 to formula (8):

$$\text{var } \hat{\alpha} = \left[\bar{x}^2 \cdot \text{var } (\hat{\beta} - \beta)\right] - \left[2\bar{x} \cdot \text{cov } (\hat{\beta} - \beta , \frac{1}{n} \textstyle\sum_{i=1}^{n} \varepsilon_i)\right] + \left[\text{var } (\frac{1}{n} \textstyle\sum_{i=1}^{n} \varepsilon_i)\right]$$

Now cov (U,V) = cov (V,U): see exercise 26.4 on page 135. So Lemma 14 proves

$$\text{cov } (\hat{\beta} - \beta , \frac{1}{n} \textstyle\sum_{i=1}^{n} \varepsilon_i) = 0$$

By the addition rule for variances (page 123),

$$\text{var } (\frac{1}{n} \textstyle\sum_{i=1}^{n} \varepsilon_i) = \frac{\sigma^2}{n}$$

Use Lemma 6:

$$\text{var } (\hat{\beta} - \beta) = \text{var } \hat{\beta} = \frac{\sigma^2}{n} \cdot \frac{1}{s_x^2}$$

So

$$\text{var } \hat{\alpha} = \frac{\sigma^2}{n} \cdot \frac{\bar{x}^2}{s_x^2} + \frac{\sigma^2}{n}$$

$$= \frac{\sigma^2}{n} \{1 + \frac{\bar{x}^2}{s_x^2}\}$$

This completes the proof.

In section 12, various estimates for σ^2 were discussed. One was

(16) $$\tilde{\sigma}^2 = \frac{1}{n}\sum_{i=1}^{n} e_i^2$$

This estimate is biased:

(17) $$E(\tilde{\sigma}^2) = \frac{n-2}{n}\cdot\sigma^2$$

To conclude this section, we will prove (17). First, in terms of the data,

(18) $$e_i = y_i - \hat{\alpha} - \hat{\beta}x_i$$

So, $\tilde{\sigma}^2$ is the MSE of the regression line. And (exercise 14.3 on page 77)

(19) $$\tilde{\sigma}^2 = \frac{1}{n}\sum_{i=1}^{n} (y_i - \bar{y})^2 - s_x^2\,\hat{\beta}^2$$

In terms of random variables,

(20) $$\tilde{\sigma}^2 = \frac{1}{n}\sum_{i=1}^{n} (Y_i - \bar{Y})^2 - s_x^2\,\hat{\beta}^2$$

Thus,

(21) $$E(\tilde{\sigma}^2) = E\{\frac{1}{n}\sum_{i=1}^{n} (Y_i - \bar{Y})^2\} - s_x^2\,E(\hat{\beta}^2)$$

In view of (5), and exercise 26.3 on page 135,

$$E(\hat{\beta}^2) = \text{var}\,\hat{\beta} + \beta^2$$

Now use Lemma 6:

$$E(\hat{\beta}^2) = \frac{\sigma^2}{n}\cdot\frac{1}{s_x^2} + \beta^2$$

So

(22) $$s_x^2\,E(\hat{\beta}^2) = \frac{\sigma^2}{n} + \beta^2\cdot s_x^2$$

We now have to work out the first term on the right side of (21). From 30.1,

$$Y_i = \alpha + \beta x_i + \varepsilon_i$$

So

$$\bar{Y} = \alpha + \beta\bar{x} + \bar{\varepsilon}, \quad \text{where} \quad \bar{\varepsilon} = \frac{1}{n}\sum_{i=1}^{n} \varepsilon_i$$

Thus

$$Y_i - \bar{Y} = \beta(x_i - \bar{x}) + (\varepsilon_i - \bar{\varepsilon})$$

And

$$\frac{1}{n}\sum_{i=1}^{n} (Y_i - \bar{Y})^2 = \beta^2 \bullet \frac{1}{n}\sum_{i=1}^{n}(x_i - \bar{x})^2 + 2\beta \bullet \frac{1}{n}\sum_{i=1}^{n} (x_i - \bar{x})(\varepsilon_i - \bar{\varepsilon})$$

$$+ \frac{1}{n}\sum_{i=1}^{n} (\varepsilon_i - \bar{\varepsilon})^2$$

Now $\frac{1}{n}\sum_{i=1}^{n} (x_i - \bar{x})^2 = s_x^2$. And $E(\varepsilon_i) = 0$, so $E(\bar{\varepsilon}) = 0$. Thus

$$E\{\frac{1}{n}\sum_{i=1}^{n} (Y_i - \bar{Y})^2\} = \beta^2 \bullet s_x^2 + E\{\frac{1}{n}\sum_{i=1}^{n} (\varepsilon_i - \bar{\varepsilon})^2\}$$

Now, as in 28.8,

$$E\{\frac{1}{n}\sum_{i=1}^{n} (\varepsilon_i - \bar{\varepsilon})^2\} = \frac{n-1}{n} \bullet \sigma^2$$

Thus

$$(23) \qquad E\{\frac{1}{n}\sum_{i=1}^{n} (Y_i - \bar{Y})^2\} = \beta^2 \bullet s_x^2 + \frac{n-1}{n} \bullet \sigma^2$$

Substitute (22) and (23) into (21):

$$E(\tilde{\sigma}^2) = \beta^2 \cdot s_x^2 + \frac{n-1}{n}\sigma^2 - \frac{\sigma^2}{n} - \beta^2 \cdot s_x^2 = \frac{n-2}{n} \cdot \sigma^2$$

This completes the proof of (17).

[There are no exercises to this section.]

32. More on box models

This section is about box models for sampling. In simple random sampling, individuals are drawn at random without replacement from some population. Thus, the draws must be made at random without replacement too. The number of draws equals the number of individuals in the sample, because the draws represent the sample. Now for the box. There is one ticket in the box for each individual in the population: the box represents the population. If there is only one data variable in the study, an individual's ticket shows the corresponding value. If, for example, there are two variables, then compound tickets are needed, showing the values of both variables.

Example. There are 4,129 currently-enrolled second-year students at a certain university. All of them took the verbal SAT at entrance. A study is made to correlate the verbal SAT with first year GPA, based on a simple random sample of 100 students. Make a box model for this study.

Solution. The box should have 4,129 tickets, one for each student in the population. Each ticket should show two numbers: the verbal SAT score, and the first-year GPA. Then 100 draws are made at random without replacement from this box. Let X_1 and Y_1 be the numbers on the first ticket -- the verbal SAT and the first-year GPA for the first student in the sample. Likewise for X_2 and Y_2, as well as X_{100} and Y_{100}. The data, consisting of 100 pairs of numbers, are to be viewed as the observed values of these random variables.

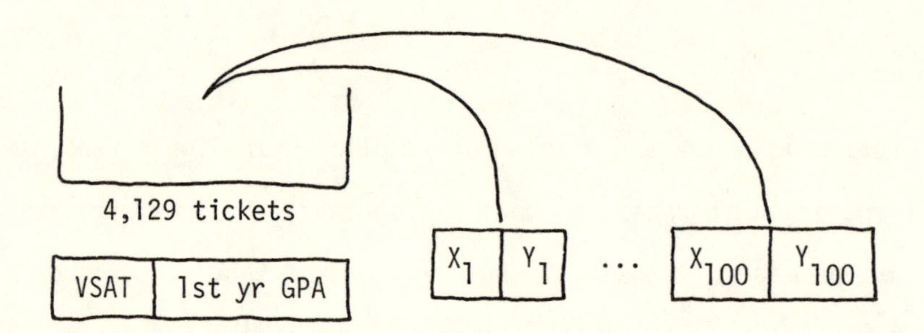

Exercise set 32

1. In the example, parameters can be defined as follows:

   μ = the average verbal SAT score for all 4,129 students

   σ = the SD of the verbal SAT scores for all 4,129 students

   ν = the average first-year GPA for all 4,129 students

   τ = the SD of first-year GPAs for all 4,129 students

   ρ = the correlation coefficient between verbal SAT score and first-year GPA for all 4,129 students

   (a) Let $\bar{X} = (X_1+\ldots+X_{100})/100$. What does $\bar{X}$ represent?

   Which parameter does it estimate?

   (b) Same, for $\bar{Y} = (Y_1+\ldots+Y_{100})/100$

   (c) Same, for $V = [\frac{1}{100}(X_1^2+\ldots+X_{100}^2)] - \bar{X}^2$

   (d) Same, for $W = [\frac{1}{100}(Y_1^2+\ldots+Y_{100}^2)] - \bar{Y}^2$

   (e) Same, for $R = \frac{1}{100}[(X_1-\bar{X})(Y_1-\bar{Y})+\ldots+(X_{100}-\bar{X})(Y_{100}-\bar{Y})]/\sqrt{VW}$

   (f) Will these estimates be exactly right? Close to right? Explain.

Notation

ν is the greek letter "nu"; τ is "tau"; ρ, "rho".

2. In a certain town, there were 10,132 single-family houses on the tax roll in 1970. A real-estate company takes a simple random sample of 100 of these houses. Their total market value in 1970 was $4,132,000. But in 1980, their total market value was $8,728,000. The ratio is $\frac{8,728,000}{4,132,000} \approx 2.11$ The company concludes that single-family houses in that town have more than doubled in price, over the period 1970 to 1980.

(a) Make a box model for this study.

(b) Is 2.11 a parameter, a random variable, or an observed value?

(c) Can the real-estate company be certain of its conclusions? Or just reasonably confident? Explain.

33. A case study

This section presents a sampling problem. The background is as follows. A company has to pay tax on increases in the value of inventory. For instance, suppose that at the beginning of the year, the company holds inventory worth \$1,000,000; at the end of the year, it holds inventory worth \$1,100,000: then it owes tax on the increase, namely \$100,000. Of course, this increase could be due to inflation, and companies are allowed to adjust for this, using LIFO.[1] In effect, the end-of-year inventory can be valued at beginning-of-year prices.

To see what this means in detail, suppose the company had complete accounting data on its end-of-year inventory, as shown in table 1. Each type of item in the inventory is given a stock number j, ranging from 1 through N: so N is the number of types of items in the inventory. For each stock number j, there is a description, the quantity q_j on hand at the end of the year (this may be zero), the price p_j at the beginning of the year, and the price p_j' at the end of the year. Due to inflation, p_j' is usually larger than p_j.

Table 1. The population. Complete LIFO accounting data.

Stock number	Description	Quantity on Hand at end of year	Price beginning of year	Price at end of year
1		q_1	p_1	p_1'
2		q_2	p_2	p_2'
.		.	.	.
.		.	.	.
.		.	.	.
N		q_N	p_N	p_N'

[1] LIFO is an acronym for "last in, first out"; the accounting convention is that the company sells first what it bought last.

Let $y_j = q_j \cdot p'_j$, the value of inventory of items of type j, at the end of the year. The total value of the inventory at the end of the year can be expressed as follows:

$$(1) \qquad v' = \sum_{j=1}^{N} y_j, \quad \text{where} \quad y_j = q_j \cdot p'_j$$

Here, the inventory has been valued at year-end prices. However, if the company elects LIFO, it can value the inventory as

$$(2) \qquad v = \sum_{j=1}^{N} x_j, \quad \text{where} \quad x_j = q_j \cdot p_j$$

This formula uses the same quantities q_j, but the lower beginning-of-year prices p_j. Thus, v is likely to be smaller then v', reducing the tax liability.

If the company has adequate computerized records, it can run off table 1, and make the computations indicated by formulas (1) and (2) without any difficulty. Many companies keep price data in a manual system, however, and the clerical labor involved in looking up two sets of prices is prohibitive. A short-cut is needed, and sampling helps.

Let $r = v'/v$. Then r is a price index: typically, $r > 1$ due to inflation. Clearly,

$$(3) \qquad v = v'/r$$

Usually, the company knows v', the value of its inventory at end-of-year prices.[1] Then, it estimates r by sampling. One possibility is to take a

[1] For instance, by taking a manual inventory, multiplying quantities on the shelf by the marked prices

simple random sample of n stock numbers.[1] For each stock number in the sample, the company can determine the quantity in stock; it can look up the end-of-year price in current records, and the beginning-of-year price in old records.

To be more specific, let j_1 be the 1st stock number in the sample. Let Q_1 be the number of items of type j_1 on hand at the end of the year.[2] Let P'_1 be the end-of-year price, and P_1 the beginning-of-year price, for this type of item. Similarly for j_2, the 2nd stock number in the sample: the quantity is Q_2, with prices P_2 and P'_2. This goes on, down to j_n, Q_n, P_n, and P'_n. The results are shown in table 2.

Table 2. LIFO accounting data on a sample of stock numbers.

Stock number	Quantity on hand at end of year	Price at beginning of year	Price at end of year
j_1	Q_1	P_1	P'_1
j_2	Q_2	P_2	P'_2
.	.	.	.
.	.	.	.
.	.	.	.
j_n	Q_n	P_n	P'_n

The company can now estimate r, using the ratio R computed from the sample:

$$(4) \qquad R = (\textstyle\sum_{i=1}^{n} Y_i)/(\sum_{i=1}^{n} X_i), \text{ where}$$

$$X_i = Q_i \cdot P_i \text{ and } Y_i = Q_i \cdot P'_i$$

[1] Stratified sampling may be preferred, although it increases the clerical labor involved by quite a lot. For more information, see the text by W. Cochran, Sampling Techniques, Wiley, New York.

[2] Formally, $Q_1 = q_{j_1}$.

The chief advantage of this method: beginning-of-year prices only have to be looked up for the relatively small number of items in the sample. Quite good accuracy can be obtained by sampling several hundred stock numbers from populations numbering in the tens of thousands.[1]

Of course, nothing is free. To draw the sample, it is necessary to make a careful list of all the types of items that may be in stock, with none left out and no duplications. The types have to be specific enough so that quantities and prices can be determined. In a grocery store, for instance, "canned tuna" is not specific enough. You have to get down to something like "6.5 ounce Star Kist #0280 Chunk Light Tuna". Drawing up the list is a fair amount of work.

Exercise set 33

1. (a) In equation (1), what does y_j represent?
   (b) Same, for x_j in equation (2).
   (c) Same for X_j in equation (4).
   (d) Same, for R in equation (4).

2. Classify each of the following as: an index of summation, a data variable, a parameter, or a random variable.

$$j, j_3, N, q_7, p, p_9', r, R$$

   Explain carefully.

[1] Technically, R is a ratio estimate; determining its accuracy is beyond the scope of this booklet. For information on this topic, see the text by W. Cochran, referenced in note 1 on page 186.

3. Where do observed values come in? Or do they?

4. Make a box model for R.

5. Compute $\frac{N}{n} E(X_1+\ldots+X_n)$.

6. The table below shows LIFO accounting data for a sample of five stock numbers. (This is real data, but the actual sample was substantially larger.[1]) Estimate the price index r. Classify each entry in the table as: a random variable, an observed value, or a parameter.

Table 3. LIFO accounting data on a sample of five stock numbers

Stock number	Quantity on hand at end of year	Price beginning of year	Price at end of year
34	16	$0.20	$0.21
144	19	0.69	0.74
189	46	1.20	1.34
288	8	.57	.60
389	44	.25	.27

(a) If X and Y are independent random variables, then $E(Y/X) = E(Y)/X$?

(b) If X and Y are independent random variables, then $\mathrm{Var}(Y/X) = (\mathrm{Var}\ Y)/X^2$? Or Var Y/Var X? Or neither?

Explain your answers.

8. In equation (4), are X_i and Y_i independent? Explain.

[1] From a study conducted by D. Freedman for C. Wood of Fairfield, California.

Answers to odd-numbered exercises

Set 1, page 2

1. $\sum_{i=1}^{3} i = 1+2+3 = 6$ 3. 14 5. 20

7. 20, it's the same sum as in exercise 5.

9. 20 11. True, by cancellation.

13. Option (ii). In (i), the squaring is omitted. In (iii), the square root does not undo the squaring, due to the intervening operation of taking the average.

15. (a) $3 \times 28 = 84$ (b) 7

17. (a) $\sum_{i=1}^{17} i$ (b) $\sum_{i=4}^{10} i^2$

Set 2, page 6

1. $3 \times 5 = 15$ 3. $2n$ 5. Marcus Welby, md

7. (a) False; j is not constant.
   (b) False; there are only $m-1$ terms.
   (c) True; there are 5 terms.

Set 3, page 8

1. (a) $\sum_{j=1}^{100} 3j = 3 \times \sum_{j=1}^{100} j = 3 \times 5{,}050 = 15{,}150$
   (b) $333{,}300 + 5{,}050 = 338{,}350$

1. (c) $j^2 = j(j-1) + j$, so it's the same as (b).

   (d) $333{,}300 + 2 \times 5{,}050 = 343{,}400$

3. (a) False; j is not a constant.

   (b) False: for instance, with two positive terms a and b,

$$(a+b)^2 = a^2 + b^2 + 2ab > a^2 + b^2$$

   (c) True.

Set 4, page 12

1. $n = 2{,}438$, the number of subjects

3. 103 mm   5. 139 lb.   7. $17,840

9. (a) True   (b) True

11. (a) True

    (b) False, the right hand expression is $(\sum_{j=1}^{m+n} u_j) + (\sum_{k=1}^{m+n} v_k)$

13. The answer to a-b-c is dollars. The answer to d-e-f is slightly bizarre: square dollars.

15. Average = 19.

Set 5, page 20

1. Both A and B are right: use (4) on page 18, with $c = 2.54$ cm/in.

3. $147 - 2 = 145$ lb.

5. Average of squared heights is 4,851.2 in^2, but the square of the average height is smaller: 4,844.16 in^2. The square of the average is smaller than the average of the squares. This will be important: see section 7, especially 7.3 on page 34.

7. (a) 68 in. (b) 156 lb. (c) 118 mm (d) 4,131

9. $5x_2$, because x_2 does not depend on i.

11. True, by 5.7 on page 19.

13. (a) average (b) m (c) y

(d) code number of subject, index of summation

15. True, by 5.7 on page 19.

17. $na = \sum_{i=1}^{n} a \leqq \sum_{i=1}^{n} x_i \leqq \sum_{i=1}^{n} b = nb$, divide by n.

Set 6, page 29

1. (a) False, $\bar{y} = -\bar{x}$ by 5.7 on page 19.

(b) True, can be argued like 6.2 on page 26.

3. Both are true: $m = \bar{x}$, use 6.1 and 6.7.

5. (a) True, see page 25.

(b) True. If SD = 0, then $\frac{1}{n} \sum_{i=1}^{n} (x_i - \bar{x})^2 = 0$; squares are never negative, so all must be 0, thus $(x_i - \bar{x})^2 = 0$ and $x_i - \bar{x} = 0$ and $x_i = \bar{x}$ for all i.

(c) False, try n = 2 and $x_1 = x_2 = 17$.

5. (d) True, because $\bar{x} = c$ and $x_i - \bar{x} = 0$ for all i.

(e) False, try $n = 2$ and $x_1 = 17$, $x_2 = -17$.

7. (a) y_j (b) $\frac{1}{6672} \sum_{j=1}^{6672} y_j$ (c) $\sum_{j=1}^{6672} y_j^2$

(d) This does not appear in the formula.

9. Let $y_i = x_i - \bar{x}$, so $z_i = \frac{1}{s} y_i$.

(a) $\bar{y} = 0$ by 5.8, so $\bar{z} = \frac{1}{s}\bar{y} = 0$ by 5.4.

(b) SD of y = SD of x = s by 6.3, so SD of z = $\frac{1}{s} \cdot$ SD of y = $\frac{1}{s} \cdot s = 1$ by 6.2.

(c) and (d) z is a pure number: the units for $x_i - \bar{x}$ and for s are the same, and cancel.

11. (a) $x_3 = 71$ in, $y_3 = 154$ lb, $x_i y_i = 10{,}934$ in × lb

(b) $\frac{1}{5} \sum_{i=1}^{5} x_i y_i = 10{,}102.6$ in × lb

$\bar{x} = 69.6$ in, $\bar{y} = 145$ lb, $\bar{x}\bar{y} = 10{,}092$ in × lb

The average of the products is different from the product of the averages. This will be important. See section 9, especially 9.5 on page 43.

13. Let $M = \frac{1}{n} \sum_{i=1}^{n} (x_i - d)^2$, the mean square deviation. So $R = \sqrt{M}$ is to be minimized. But the d which makes M smallest also makes R smallest. By exercise 12,

$$M = [\frac{1}{n} \sum_{i=1}^{n} (x_i - \bar{x})^2] + (\bar{x} - d)^2$$

The term $(\bar{x} - d)^2$ is 0 or positive; so M is smallest when $\bar{x} = d$, and then

$$R = \sqrt{M} = \text{SD of } x$$

15. Put half the x's at a, the other half at b, so $\bar{x} = \frac{1}{2}(a+b)$ and $SD = \frac{1}{2}(b-a)$.

Set 7, page 35

1. (a) True, see 7.2 (b) False
   (c) True, see 7.2 (d) False

3. $SD = \sqrt{var} = \sqrt{20} \approx 4.5$ cm, so it's (ii).

5. 9 square years. The units of variance are bizarre.

7. $\bar{z} = c\bar{x} + d$, so $z_i - \bar{z} = c(x_i - \bar{x})$ and $\text{var } z = \frac{1}{n}\sum_{i=1}^{n} (z_i - \bar{z})^2$ $= \frac{1}{n}\sum_{i=1}^{n} c^2(x_i - \bar{x})^2 = c^2 \frac{1}{n}\sum_{i=1}^{n} (x_i - \bar{x})^2 = c^2 \text{ var } x$. This is the same as exercise 6.8.

9. False, multiply by $2^2 = 4$.

11. Then the variance is 0, by formula 3 on page 34. So the SD is 0 too, and the variable must be constant: see exercise 6.5 on page 30.

Set 8, page 41

1. (a) $\mu = 53$ in and $\sigma = 2$ in (b) $N(53,4)$

3. mean = 16 and SD = 5

Set 9, page 44

1. Positive, taller people tend to be heavier.

3. var ht ≈ 8.9 in^2, var wt ≈ 406 lb^2
   cov(ht,wt) ≈ 0.56 in $\times$ lb

5. $\mathrm{cov}(ht,inc) = 1{,}179{,}500 - 69 \times 17{,}000$

$= 6{,}500$ inch × dollar

7. (a) n (b) $\bar{x}$ (c) $y_i - \bar{y}$ (d) $\sum_{i=1}^{n} (x_i - \bar{x})(y_i - \bar{y})$

9. $(x_i - \bar{x})(y_i - \bar{y}) = (y_i - \bar{y})(x_i - \bar{x})$, add and divide by n.

11. $\bar{z} = m\bar{x} + b$, so $z_i - \bar{z} = m(x_i - \bar{x})$ and $\frac{1}{n}\sum_{i=1}^{n} (z_i - \bar{z})(u_i - \bar{u}) = m \cdot \frac{1}{n}\sum_{i=1}^{n} (x_i - \bar{x})(u_i - \bar{u})$

13. $\bar{z} = a\bar{u} + b\bar{v} + c$, so $z_i - \bar{z} = a(u_i - \bar{u}) + b(v_i - \bar{v})$ and $(z_i - \bar{z})^2 = a^2(u_i - \bar{u})^2 + 2ab(u_i - \bar{u})(v_i - \bar{v}) + b^2(v_i - \bar{v})^2$. Add and divide by n.

15. (a) True, n is used as an index of summation; this is non-standard, but OK.

(b) False, because the number of subjects is m, not n.

Set 10, page 51

1. (i) is ok; the numerator is $\mathrm{cov}(x,y)$; the denominator is $\sqrt{\mathrm{var}\ x \cdot \mathrm{var}\ y} = \sqrt{\mathrm{var}\ x} \cdot \sqrt{\mathrm{var}\ y}$. See formula 10.4.

(ii) is ok; multiply (i) by n in numerator and denominator.

(iii) is not ok; there are n's missing in front of $\bar{x}\,\bar{y}$, $\bar{x}^2$, and $\bar{y}^2$.

3. (a) same, see formula 10.6.

(b) same, see formula 10.7.

5. Sure: cov(length, width) = $0.42 \times 7 \times 12\ \mathrm{ft}^2 = 35.28\ \mathrm{ft}^2$ by formula 10.3, so average area = $35.28 + 53 \times 107 = 5{,}706.28\ \mathrm{ft}^2$ by proposition 9.2, and total area = $89 \times 5{,}706.28 \approx 507{,}859\ \mathrm{ft}^2$.

7. a) x_i b) $x_i - \bar{x}$ c) not there d) $\sum_{i=1}^{6,672} y_i^2$ e) not there

f) $\sqrt{\frac{1}{6,672}\sum_{i=1}^{6,672}(x_i-\bar{x})^2}$ g) $(\frac{1}{6,672}\sum_{i=1}^{6,672} y_i^2) - (\bar{y}^2)$

9. r is undefined, see page 51.

Set 11, page 58

1. Ave ht = 28,359/409 = 69.33741 in.

   Ave (ht^2) = 1,969,716/409 = 4815.93154 in^2.

   var ht = 4815.93154 - $(69.33741)^2$ = 8.25511 in^2, by 7.3.

   ave wt = 158.77262 lbs.

   ave (ht x wt) = 4,513,810/409 = 11,036.21027 in x lb.

   cov (ht, wt) = 11,036.21027 - (69.33741 x 158.77262), by 9.2.

   = 27.32802 in x lb

   slope = 27.32802/8.25511 = 3.31044 lb/in, by 11.5

   intercept = 158.77262 - (3.31044 x 69.33741), by 11.3.

   = -70.76472

   Rounding off a bit, the line is

   predicted weight = (3.3 lbs/in) x height - 71 lbs

   The arithmetic is quite annoying; you have to keep many decimals in the intermediate steps to get accuracy to one or two decimals in the final answer. In working such problems, do not worry about minor round-off errors. And do not give your final answer with many meaningless decimal places!

3. a) Measurement error

   b) predicted length = (0.0522 cm/kg) load + 287.115 cm

   c)

Load (kg)	2	3	5	105
pred. length (cm)	287.22	287.27	287.38	???

A load of 105 kg is so far from the data that the regression equation cannot be trusted. For example, the spring might break under such a heavy load.

d) The regression equation is probably better, because it combines information from 8 data points, reducing the effect of measurement error.

e) intercept = 287.115 cm; this is probably more trustworthy than the observation in the table, as in d).

f) slope = .0522 cm/kg

5. The numerator is cov(x, y) by exercise 9.12; the denominator is var x ; now refer to 11.5.

7. a) $\frac{1}{3,091} \sum_{i=1}^{3,091} (x_i - \bar{x})^2$

b) Not there

c) $\frac{1}{3,091} \sum_{i=1}^{3,091} (x_i - \bar{x})(y_i - \bar{y})$

d) Not there

9. True, because minimizing the r.m.s. error is the same as minimizing the MSE:

$$\text{r.m.s. error} = \sqrt{\text{MSE}}$$

11. The two lines are different:

$$\text{predicted } y = \text{intercept} + r \cdot \frac{s_y}{s_x} \cdot x$$

$$\text{predicted } x = \text{intercept} + r \cdot \frac{s_x}{s_y} \cdot y$$

The two r's are the same, because r(x, y) = r(y, x) . The ratio of the slopes is var y/var x .

Set 12, page 64

1. Differentiation with respect to d gives $-2\frac{1}{n}\sum_{i=1}^{n}(x_i-d) = 0$, so $\frac{1}{n}\sum_{i=1}^{n}(x_i-d) = 0$, so $(\frac{1}{n}\sum_{i=1}^{n}x_i) - d = 0$, so $d = \frac{1}{n}\sum_{i=1}^{n}x_i$. That is, d is the average. Compare exercise 6.13.

Set 14, page 76

1. False, the squares are missing. In the height-weight example,

   SD of y = 25 lb (given),

   SD of $\hat{y} = r\cdot$(SD of y) = 0.36×25 lb = 9 lb, by formula 14.9,

   SD of $e = \sqrt{1-(.36)^2} \times 25 \approx 23.32$ lb, by formula 14.11,

   and

$$25^2 = 9^2 + (23.32)^2 \text{ , this is formula 14.4}$$

$$25 \neq 18 + 23.32$$

3. Use 14.4 and 14.8.

Set 15, page 81

1. (a) True, this is just formula 14.4 in other language

   (b) False, this is just exercise 14.1 in other language

3. 98%, 14%. Note that 98% + 14% ≠ 100%. Work: $r^2 \approx .98$, $\sqrt{1-r^2} \approx .14$

5. 13%. Work: $r^2 \approx .13$; switching height and weight does not change r, by formula 10.6.

Set 17, page 91

1. The random variable is the procedure; the observed value, 2.

3. True. 5. X goes with 4, Y with 3, but not for sure.

Set 18, page 94

1. a) Draw a ticket at random from the box, take the product of the two numbers on the ticket.

   b) 2/4 = 1/2 c) 1/4

Set 19, page 96

1. The two random variables are independent.

3. The two are dependent: if X is 1, then Y is 1 or 3; if X is 7, then Y must be 11.

5. The two are independent.

7. True.

Set 20, page 101

1. a) 1/4 b) 1/4 c) 2/4 = 1/2

3. a) 1/3 b) 2/3 c) 1/3 d) no e) no 5. Without

Set 22, page 108

1. Values of Z

X \ Y	0	1
1	3	7
2	6	10
2	6	10

Distribution table for Z

value	chance
3	1/6
6	2/6
7	1/6
10	2/6

3. Values of S

first draw \ second draw	1	2	3	4	5	6
1	x	3	4	5	6	7
2	3	x	5	6	7	8
3	4	5	x	7	8	9
4	5	6	7	x	9	10
5	6	7	8	9	x	11
6	7	8	9	10	11	x

Distribution table for X

value	chance
3	2/30
4	2/30
5	4/30
6	4/30
7	6/30
8	4/30
9	4/30
10	2/30
11	2/30

5. a) No, X and Y are dependent: if X is 1 then Y can be 1, and if X is 3 then Y cannot be 1.

b) Distribution table for 2X - Y

value	chance
0	1/4
1	1/4
4	1/4
6	1/4

7. The box for U is [3 | 6 | 7], so U is better.

9. 3 goes with V , and 5 with U , for sure.

Set 23, page 114

1. a) false b) false c) false d) false e) false
f) true g) false h) false i) true

3. The chances add up to 1, so the missing chance is 2/10, and

$$E = 1 \times \frac{3}{10} + 2 \times \frac{1}{10} + 3 \times \frac{2}{10} + 7 \times \frac{4}{10} = \frac{39}{10} = 3.9$$

Set 24, page 118

1. a) $E(U) = \frac{1+7+6+2}{4} = 4$

   b) yes

   c) $E(V) = \frac{0+10+9+0}{4} = 4\frac{3}{4}$

   d) No

3. a) $E(W) = E(X \cdot Y) = E(X) \cdot E(Y) = \frac{9}{4} \cdot 2 = 4\frac{1}{2}$

   b) W stands for this chance procedure: draw a ticket at random from the first box; independently, draw a ticket at random from the second box; take the product of the numbers on the two tickets.

5. a) True b) False, $E(x)$ is meaningless, as is $ax + bY$

   c) False or meaningless d) True

   e) False, unless you assume independence.

Set 25, page 127

1. 1000 ± 10 or so: $SE = \sqrt{var} = \sqrt{100}$.

3. a) True b) False c) True d) False e) False f) True

5. a) Y stands for this chance procedure: draw a ticket at random from the box, multiply the number on it by 10.

   b) $10 \cdot$ SE of X .

   c) $100 \cdot$ var X .

   d) 2/3

   It is important to see the difference between exercises 4 and 5. In exercise 4, we added up 10 independent draws. In exercise 5, we multiplied one draw by 10.

7. var $X = \frac{14}{3}$, var $Y = \frac{24}{3}$, var $(X + Y) = \frac{62}{3} \neq$ var X + var Y , due to dependence.

9. The list is static; it is there, and involves no uncertainty, no chances. The random variable is dynamic: it will generate a number, and we don't know which one. Uncertainty is involved, and chance.

Set 26, page 135

1. $E\{X - E(X)\} = E(X) - E(X) = 0$.

3. $[X - E(X)]^2 = X^2 - 2E(X) \cdot X + E(X)^2$

$E\{[X - E(X)]^2\} = E(X^2) - 2E(X) \cdot E(X) + E(X)^2 = E(X^2) - E(X)^2$

5. $E[(X + Y)Z] = E(XZ) + E(YZ)$

$E[(X + Y)] \cdot E(Z) = E(X) \cdot E(Z) + E(Y) \cdot E(Z)$

Subtract.

7. No, $\text{cov}(X, Y) \neq 0$

Set 27, page 142

1. Your net gain is like the sum of 1000 draws made at random with replacement from the box

\$35	37 -\$1

average of box = -\$1/19 and SD ≈ \$5.76. So expected value = 1000 x - \$1/19 ≈ -\$53 and SE = $\sqrt{1000}$ x \$5.76 ≈ \$182. You will lose around \$53, give or take \$182 or so. There is lots of room for chance error here -- which is why people gamble.

3. No. If 0 or 00 or any odd number comes up, both lose \$18. If (say) 24 comes up, then A wins \$35 on 24, and loses \$17 on the remaining numbers, for a net gain of \$35 - \$17 = \$18. And B wins \$18 too.

5. net gain of gambler A = $X_1 + \dots + X_{10}$

net gain of gambler B = $10X_1$

net gain of gambler C = $X_1 + \dots + X_5$

Set 28, page 151

1. a) 3.7 ± 0.2 or so

b) i) observed value ii) parameter iii) observed value

iv) parameter

3. a) False b) True c) True

5. False, in the formula $\text{var } \bar{X} = \sigma^2/n$, we are thinking of $\bar{X}$ as a random variable, not data. See page 127.

7. a) $E(X_1)$ and $\text{var } X_1$ are parameters, the other two are random variables

b) Both are false: $E(\bar{X})$ is the expected value of the random variable $\bar{X}$, while $\frac{1}{n}\sum_{i=1}^{n} X_i$ expresses $\bar{X}$ in terms of the individual draws. Similarly for $\text{var } \bar{X}$.

Set 29, page 158

1. 403 ± 1

3. $e_3 = 406 - 403 = 3$, but ε_3 is not observable.

5. False, see exercises 28.5 and 28.7.

Set 30, page 170

1. The regression line was found in example 1:

$$y = .049x + 439.011$$

For the third data point, from table 2

$$x_3 = 4 \text{ and } y_3 = 439.21$$

So the predicted length is

$$.049 \times 4 + 439.011 = 439.207$$

And

$$e_3 = 439.21 - 439.207 = .003 \text{ cm}$$

However, ε_3 is not observable.

3. $\hat{\alpha}$ = estimated unloaded length = 287.115 cm

$\hat{\beta}$ = estimated constant of proportionality = .052 cm/kg.

These are observed values.

$\hat{\sigma} \approx 1.1547 \times .0903 \times .1879 \approx .0196$ eqn (12)

SE of $\hat{\alpha} \approx .007 \times 1.472 \approx .010$ cm eqn (8)

SE of $\hat{\beta} \approx .007 \times .279 \approx .002$ cm/kg eqn (9)

5. No, see exercise 28.7b.

7. We are thinking of $\hat{\alpha}$ as a random variable -- to compute its SE; it is observable.
σ is an unknown parameter.
n, $\bar{x}$, and s_x are known parameters.

Set 32, page 182

1. a) $\bar{X}$ is the average VSAT score for the 100 sample students, and it estimates μ

   b) $\bar{Y}$ is the average 1st-year GPA for the 100 sample students, and it estimates ν

   c) V is the variance of the VSAT scores for the 100 sample students, and it estimates σ^2

   d) W is the variance of the 1st-year GPAs for the 100 sample students, and it estimates τ^2

   e) R is the correlation coefficient between VSAT scores and 1st-year GPAs among the 100 sample students, and it estimates ρ.

   f) These estimates will be close to right, but will be off by small chance errors.

Set 33, page 187

1. a) y_j is the value of inventory of items of type j, at year-end prices.

   b) x_j is the value of inventory of items of type j, at beginning-of-year prices.

   c) For the ith item-type in the sample, the value of the inventory at beginning-of-year prices is X_i

   d) R is the price index based on the sample: the value at end-of-year prices divided by the value at beginning-of-year prices.

3. They come in after you have the sample.

5. Value of inventory at beginning-of-year prices.

7. a) False.

   b) Neither.